"*Blockchain Technology: Transforming Businesses and Shaping the Future* is a wonderful resource for those seeking to deepen their understanding of blockchain technology. With the field rapidly evolving, this book offers a solid foundation in the fundamentals while also presenting advanced research for seasoned practitioners and scholars. Whether you're a student in the classroom or a professional in the office, this comprehensive guide serves as an essential reference, illuminating the potential of blockchain across industries and the supply chain."

–Professor Joseph Sarkis, Worcester Polytechnic Institute, USA

"*Blockchain Technology: Transforming Businesses and Shaping the Future* is an invaluable resource for readers from various backgrounds and disciplines who wish to deepen their understanding of blockchain technology. It combines both theoretical and practical perspectives, thoroughly exploring blockchain practices and challenges across multiple sectors. Authored by renowned experts, the book provides critical insights into the design, management, and control of blockchain technology for businesses."

–Professor CT Lin, Distinguished Professor of Artificial Intelligence and Computer Science, University Technology Sydney, Australia

"The growth of research on supply chain digitization in general, and blockchain technology in particular, has been immense. At the same time, the practical application of blockchain technology is still in its infancy. Therefore, a compendium on when and how to use blockchain technology in business and supply chains, coupled with business cases and examples, is a rich source for practitioners and academics alike. The book summarizes the current state of knowledge on blockchain technology and offers a critical assessment of the technology's application areas – from supply chain traceability to gold trading. What I appreciate most is that after reading the book, it becomes clear that successful blockchain projects are not just about the technology."

–Stephan M. Wagner, Professor of Supply Chain Management, ETH Zurich

"Blockchain is a major technology which commands a significant and disruptive impact on contemporary supply chains and on businesses operating all over the world and in numerous sectors. This book will enhance and support a holistic and detailed understanding of the role, impact and potential of the Blockchain technology, and it will be an excellent source for academic scholars, managers, practitioners and students. I am confident that it will also pave the way for other books to follow in this emerging and important field of study."

–Professor Michael Bourlakis, Centre of Logistics, Procurement and Supply Chain Management, Cranfield School of Management, United Kingdom

"The book *Blockchain Technology: Transforming Businesses and Shaping the Future* comprehensively presents principles and technologies for management and implementation of blockchain in real life. It both offers theoretical foundations on Blockchain and presents numerous real-world applications. The chapters are written by well-known experts and organized in a coherent way covering major technological concepts, supply chain management, and a variety of industry case studies. This book will serve as a valuable resource for professionals and academics alike."

–Prof. Dr. habil. Dmitry Ivanov, Professor of Supply Chain and Operations Management at Berlin School of Economics and Law, Germany

"The book *Blockchain Technology: Transforming Businesses and Shaping the Future* is an excellent resource for anyone interested in understanding the key opportunities, challenges, and implementation of blockchain technology for businesses. This timely book provides fresh insights into a critical aspect of business and supply chain management through the lens of blockchain technology. Effective process innovation, including the deployment of blockchain technology, is key to future success. The book offers both theoretical and practical perspectives, along with up-to-date critical insights on the design and management of blockchain technology for businesses."

–Prof. Edward Sweeney, Professor of Logistics and Supply Chain Management and Deputy Executive Dean, School of Social Sciences and Edinburgh Business School, Heriot-Watt University

"Blockchain applications are evolving quickly, and this book will bring you up to date with the latest research and thinking on business opportunities and solutions. The myriad applications include healthcare, agriculture, supply chains, and international trade. Each chapter is well-referenced, pointing to even further sources, making this a helpful resource for practitioners and scholars."

–Prof. Tyson R. Browning, Professor of Operations Management, Neeley School of Business, Texas Christian University, USA

Blockchain Technology

Blockchain technology is considered a disruptive innovation that changes the ways companies and global processes operate. This technology has impressive powers to change this world for the better.

This book examines the origins, emergence, challenges, and opportunities in the blockchain field, rethinking business strategy and readiness in the digital world and how blockchain technology would improve businesses. It provides a blockchain readiness model for managing supply chains and reviews enabling technologies such as AI, big data, and organisational capabilities that support the adoption of blockchain technology. Through innovative design and simulation of a blockchain framework, it aims to enhance the traceability and transparency of business operations and supply chains. This includes developing key performance indicators for measuring the seamless integration of blockchain technology and achieving a successful outcome. It explores how blockchain technology enhances the green and sustainability aspects of businesses by comparing the sectors and discussing the potential for blockchain to promote a green and sustainable economy. This book concludes with research frontiers and blockchain applications in healthcare, international trade, and supply chain sectors.

Key features

- Integrates both theoretical and practical perspectives.
- Includes material that is informative for readers from diverse backgrounds and disciplines.
- Explores blockchain technology practices and challenges in-depth across various sectors.
- Offers up-to-date, critical insights on the design, management, and control of blockchain technology for businesses.
- Written by experts with extensive experience in the field.

It is primarily written for senior undergraduate and graduate students and academic researchers in fields including buisness management, industrial and systems engineering, supply chain, electrical engineering, electronics and communication engineering, computer engineering, and information technology.

Smart Technologies for Engineers and Scientists
Series Editor:
Mangey Ram

Smart Technologies including artificial intelligence, the Internet of Things (IoT), machine learning, and big data computing have played an important role in the advancement of processes and the functioning of machines. Smart systems are being used in a wide range of industries such as transportation, energy, healthcare, safety and security, logistics, and manufacturing. This series discusses the use of smart technologies in diverse areas of research including electric vehicles, advanced manufacturing techniques, Industry 4.0, and nanoelectronics.

Blockchain Technology: Transforming Businesses and Shaping the Future
Edited by Jay Daniel, Ashutosh Samadhiya, Jose Arturo Garza-Reyes

Practical Antenna: Design, Analysis, and Applications
Edited by Abhishek Kumar Awasthi, Abhinav Sharma, Kushmanda Saurav

Sustainable Energy Solutions with Artificial Intelligence, Blockchain Technology, and Internet of Things
Edited by Arpit Jain, Abhinav Sharma, Vibhu Jately, Brian Azzopardi

Internet of Things in Modern Computing: Theory and Applications
Edited By Vinay Chowdary, Abhinav Sharma, Naveen Kumar, Vivek Kaundal

Applications of Mathematical Modeling, Machine Learning, and Intelligent Computing for Industrial Development
Edited by Madhu Jain, Dinesh K. Sharma, Rakhee Kulshrestha, and H.S. Hota

Advances in Mathematical and Computational Modeling of Engineering Systems
Edited by Mukesh Kumar Awasthi, Maitri Verma, Mangey Ram

For more information about this series, please visit: https://www.routledge.com/Smart-Technologies-for-Engineers-and-Scientists/book-series/CRCSTFES

Blockchain Technology
Transforming Businesses and Shaping the Future

Edited by
Jay Daniel, Ashutosh Samadhiya, and
Jose Arturo Garza-Reyes

CRC Press is an imprint of the
Taylor & Francis Group, an **informa** business

Front cover image: greenbutterfly/Shutterstock

First edition published 2025
by CRC Press
2385 NW Executive Center Drive, Suite 320, Boca Raton FL 33431

and by CRC Press
4 Park Square, Milton Park, Abingdon, Oxon, OX14 4RN

CRC Press is an imprint of Taylor & Francis Group, LLC

© 2025 selection and editorial matter, Jay Daniel, Ashutosh Samadhiya and Jose Arturo Garza-Reyes]; individual chapters, the contributors

Reasonable efforts have been made to publish reliable data and information, but the author and publisher cannot assume responsibility for the validity of all materials or the consequences of their use. The authors and publishers have attempted to trace the copyright holders of all material reproduced in this publication and apologize to copyright holders if permission to publish in this form has not been obtained. If any copyright material has not been acknowledged please write and let us know so we may rectify in any future reprint.

Except as permitted under U.S. Copyright Law, no part of this book may be reprinted, reproduced, transmitted, or utilized in any form by any electronic, mechanical, or other means, now known or hereafter invented, including photocopying, microfilming, and recording, or in any information storage or retrieval system, without written permission from the publishers.

For permission to photocopy or use material electronically from this work, access www.copyright.com or contact the Copyright Clearance Center, Inc. (CCC), 222 Rosewood Drive, Danvers, MA 01923, 978-750-8400. For works that are not available on CCC please contact mpkbookspermissions@tandf.co.uk

Trademark notice: Product or corporate names may be trademarks or registered trademarks and are used only for identification and explanation without intent to infringe.

ISBN: 978-1-032-61473-1 (hbk)
ISBN: 978-1-032-89428-7 (pbk)
ISBN: 978-1-003-54276-6 (ebk)

DOI: 10.1201/9781003542766

Typeset in Sabon
by codeMantra

Contents

About the book	viii
Preface	ix
About the editors	xii
Acknowledgements	xv
List of contributors	xvi

PART 1
Introduction

1 Mapping the knowledge domain of blockchain technology studies: A scientometric analysis   3
JAY DANIEL AND ANDREW STAPLETON

PART 2
Rethinking strategy and blockchain readiness in the digital world

2 Blockchain technology readiness model for supply chain management   33
AHMED ALMAAZMI AND JAY DANIEL

3 Modeling of blockchain interruptions for smooth integration with supply chain: An ISM-MICMAC mapping approach   49
SACHIN YADAV AND SHUBHANGINI RAJPUT

4 Enablers of blockchain adoption in organisations: A view of digital assets and organisational capabilities   66
ANDREW HIRST, CHRISTIAN MICHAEL VEASEY, AND JAY DANIEL

PART 3
Blockchain adoption and enabling technologies

5 Blockchain technology and artificial intelligence for enhanced vaccine supply chain management   89

UTKARSH MITTAL AND AMIT KUMAR YADAV

6 Harnessing synergy: A multidimensional exploration of AI, big data, and blockchain convergence in the healthcare sector   105

SREEJITH BALASUBRAMANIAN, VINAYA SHUKLA, SHALINI AJAYAN, SONY SREEJITH, AND ARVIND UPADHYAY

7 An innovative design and simulation of a blockchain-based smart contract framework for enhancing gold traceability   129

DEO SHAO, MARTHA SHAKA, FREDRICK ISHENGOMA, GEORGE BENNETT, FREDRICK BETUEL SAWE, AND JAY DANIEL

PART 4
Managing the blockchain technology

8 Managing and planning blockchain implementation framework: A conceptual model   151

JAY DANIEL AND JEHAN ZAIB

9 Ensuring transparency: The blockchain impact on agriculture and livestock traceability   178

RAKESH G NAIR, PANKAJ KUMAR DETWAL, AND M. MUTHUKUMAR

10 The role of blockchain in the sustainability of green technology   198

RAGESHREE SINHA, SUSHIL MOHAN, AND ARVIND UPADHYAY

PART 5
Blockchain technology and practices in business management

11 Blockchain applications in healthcare sector: Opportunities and challenges   217
BISWAKSEN MISHRA AND RAJESH KUMAR SINGH

12 Improving transparency and efficiency in international trade through blockchain technology   240
WITOLD BAHR, LEE EE YERN, AND IAN MCEWAN

Index   259

About the book

Blockchain technology is considered a disruptive innovation that changes the ways companies and global processes operate. This technology has impressive powers to change this world for the better. However, some obstacles to its adoption have to be addressed. This book aims to assist academics, researchers, practitioners, and students in adopting blockchain technology to drive business and supply chain transformation and develop blockchain models and practices within their supply chains and businesses. This book examines the origins, emergence, challenges, and opportunities in the blockchain field, rethinking business strategy and readiness in the digital world and how blockchain technology would improve businesses. It provides a blockchain readiness model for managing supply chains and reviews enabling technologies such as AI, big data, and organisational capabilities that support the adoption of blockchain technology. Through innovative design and simulation of a blockchain framework, it aims to enhance the traceability and transparency of business operations and supply chains. This includes developing key performance indicators for measuring the seamless integration of blockchain technology and achieving a successful outcome. It explores how blockchain technology enhances the green and sustainability aspects of businesses by comparing the sectors and discussing the potential for blockchain to promote a green and sustainable economy. This book concludes with research frontiers and blockchain applications in healthcare, international trade, and supply chain sectors.

Preface

Blockchain technology is a distributive innovation and impactful megatrends reshaping businesses, industries, and commerce. Blockchain technology is influencing how we collaborate and exchange information across businesses and supply chains, as well as how we integrate, manage, and control business and supply chain operations. Blockchain technology has impressive potential to positively transform businesses and the world for the better. Blockchain aims to enhance the traceability and transparency of business operations and supply chains by transforming the design, management, and control of contemporary businesses and shaping future business and supply chain practices.

In Chapter 1, Daniel and Stapleton provide an overview of blockchain technology studies including the origins, emergence, challenges, and opportunities in the blockchain field. This chapter evaluates research trends, performance, and state-of-the-art literature through bibliometric and scientometric analyses in the blockchain field. It demonstrates hot topics and emerging trends recognised in this area encompassing blockchain and security, smart contracts, challenges, privacy, management, the Internet of Things (IoT), Bitcoin, framework development, and supply chain management. This chapter reveals several challenges, including issues related to data protection and privacy, the seamless interoperability and integration of IoT and other digital technologies with blockchain, the sustainability of supply chains, and the assurance of transparency and traceability in products and services. Furthermore, solving scalability issues remains a paramount concern within the blockchain domain.

The second part of this book, Chapters 2–4, develops, analyses, and critically appraises blockchain readiness and rethinking strategy in the digital world. In Chapter 2, Almaazmi and Daniel propose a blockchain technology readiness model for supply chain management. They identify key variables affecting blockchain readiness for supply chain management such as leadership support, infrastructure, regulations, and collaboration. The model would be a cornerstone for private and public organisations to adopt blockchain technology. In Chapter 3, Yadav and Rajput identify and study the barriers to integrating blockchain technology in startup supply chains and identify the influential barriers through an interpretive structural modelling (ISM) mapping approach. In Chapter 4, Hirst, Veasey, and Daniel review the enablers of blockchain adoption focusing on digital assets and organisational capabilities. They propose a framework for enablers of blockchain adoption in organisations and discuss the critical enablers of blockchain technology and the role that digital assets, leadership, and strategy play in the development of blockchain.

The third part of this book, Chapters 5–7, reviews, analyses, and designs blockchain adoption and enabling technologies such as AI, machine learning, simulation, IoT, and big data. In Chapter 5, Mittal and Yadav propose an intelligent vaccine supply management system that integrates blockchain technology, IoT sensors, and machine learning algorithms. The blockchain architecture fosters transparency, coordination, and performance across the entire vaccine production and supply chain. In Chapter 6, Balasubramanian, Shukla, Ajayan, Sreejith, and Upadhyay review a multidimensional exploration of AI, big data, and blockchain convergence in the healthcare sector. They introduce four key dimensions: patient centricity, research and development, integration and collaboration, and the creation of intelligent environments that play a critical role in healthcare delivery. In Chapter 7, Shao, Shaka, Ishengoma, Bennett, Sawe, and Daniel design an innovative simulation of a blockchain framework for enhancing gold mining traceability. They develop and implement the blockchain framework through a GoldNet showcasing the traceability of a gold supply chain from its extraction to retail. The proposed GoldNet framework can potentially improve the transparency of gold trading systems.

The fourth part of this book, Chapters 8–10, addresses opportunities and challenges in the management of blockchain technology. In Chapter 8, Daniel and Zaib propose the management and planning of blockchain implementation framework, which provides a comprehensive framework for the seamless integration of blockchain technology to achieve a successful outcome. The comprehensive framework can assist organisations with effectively managing, implementing, and evaluating the performance of blockchain technology in practice. In Chapter 9, Nair, Detwal, and Muthukumar explore and integrate blockchain technology into the agri-food supply chain, enabling traceability and transparency in the agriculture and livestock industry. They develop a blockchain-enabled supply chain model for a livestock supply chain. In Chapter 10, Sinha, Mohan, and Upadhyay review the role of blockchain in the sustainability of green technology. They highlight the transforming capabilities of blockchain technology, which can create more efficiency in achieving a net-zero outcome to tackle the climate emergency through knowledge sharing and collaboration.

The final part of this book, Chapters 11 and 12, presents blockchain applications in healthcare and international trade along with opportunities and challenges. In Chapter 11, Mishra and Singh review blockchain applications in the healthcare sector and discuss challenges and opportunities such as electronic healthcare records and early diagnosis and treatment of life-threatening diseases. In Chapter 12, the concluding chapter, Bahr, Yern, and McEwan examine improving transparency and efficiency in international trade through blockchain technology. They demonstrate applications of blockchain technology in various aspects of global trade, including supply chains, trade finance, and customs clearance, along with their corresponding challenges and opportunities.

Transforming businesses, industries, and supply chains to adopt and utilise blockchain technology will result in disruptive change across many sectors. The transformation presents formidable challenges, but blockchain and digital technologies are already having very significant effects in reengineering and rearchitecting business and supply chains. The studies reported in this book provide insights and analysis on the impact of blockchain technology across the business and supply chain landscapes of many sectors, citing the latest and most seminal work throughout. The digitalisation of business, commerce, and industry will affect supply chains, supply networks, and business ecosystems in diverse ways across different industries and sectors. This book provides academics, researchers, practitioners, and students in adopting blockchain technology to drive business and supply chain transformation and develop blockchain models and practices within their supply chains and businesses.

Jay Daniel

Ashutosh Samadhiya

Jose Arturo Garza-Reyes

About the editors

Dr Jay Daniel is an associate professor of digital supply chain and technology innovation research cluster lead at the Centre for Supply Chain Improvement at the University of Derby, UK. Before joining the Derby Business School, he was a lecturer in Supply Chain and Information Systems at the University of Technology Sydney (UTS), Australia. Previously with DB Schenker, Australia, and Alliance International Registrar, Asia Pacific, he held positions of Senior Management Consultant, Supply Chain Solution Analyst, Project Manager, Industry Trainer, and Lead Auditor. He has published over 60 refereed papers in many prestigious journals and international conferences such as *IEEE Transactions on Fuzzy Systems, International Journal of Information Management, Service Research, and Innovation*, amongst others. He is serving on the editorial board of several top journals such as *Australasian Journal of Information Systems* (ERA rank A), *Quality Management Journal*, and *Operations Management Research* and on several leading scientific/program committees of national/international conferences. His research is funded by government agencies such as Economic and Social Research Council, InterAct, Made Smarter Innovation, UK Research and Innovation and industry collaborators around the world. He has been invited as a keynote speaker/invited speaker at international industry and academic workshops and conferences around the globe such as Keynote Speaker in Oracle Modern Business Experience Conference (Australia), and World-class speaking about sustainable Supply chain at IISRP (USA, France, Italy, China, Korea, Russia, Germany, Japan, Austria, etc.). He is honoured to be included in the Top 50 Thought Leaders and Influencers on Supply Chain developed by thinkers 360.

Ashutosh Samadhiya is a distinguished academic figure in the field of Operations and Supply Chain Management, known for his significant contributions to research and education. Currently, he is serving as an Assistant Professor at Jindal Global Business School, OP Jindal Global University, Sonipat, India. He has amassed over eight years of experience in teaching, research, and innovation. His academic journey culminated in the attainment of a Ph.D. in Operations and Supply Chain Management from the esteemed Indian Institute of Technology Roorkee (IIT Roorkee), India. During his doctoral studies, he showcased his innovative prowess by developing a groundbreaking handloom machine, a feat that earned him an Indian patent, granted by the Indian Government. Furthermore, he demonstrated his commitment to technological advancement by transferring the technology associated with the handloom machine. In recognition of his academic excellence and research acumen, he was selected for the prestigious Newton-Bhabha PhD placement program. This opportunity led him to collaborate with the Centre of Supply Chain at the University of Derby, UK, in 2021. Since then, he has maintained an honorary visiting researcher position at the University of Derby, fostering international collaborations and enriching his academic pursuits. His scholarly endeavours extend beyond academia, as evidenced by his prolific publication record in international referred and national journals, as well as presentations at renowned conferences on both the international and national stages. His research interests span a wide array of topics, including sustainability science, green/sustainable supply chain management, green operations, sustainable procurement, sustainable development, circular economy, Industry 4.0, performance measurement, human capital in supply chain and operations, decision modelling for sustainable business, and integration of operational areas with other domains. Moreover, he possesses a diverse skill set, encompassing advanced statistical models, multivariate analysis, multi-criteria decision-making, fuzzy theory, fuzzy optimisation, fuzzy multi-criteria decision-making, grey theory and analysis, machine learning, and more. His proficiency in these areas equips him to undertake complex research projects across various settings, including rural, semi-urban, and urban environments. In addition to his scholarly achievements, he holds a design patent for an artificial intelligence cash counting machine, showcasing his innovative spirit and commitment to technological innovation. Furthermore, he has made significant contributions to the academic community by serving as a reviewer and guest editor for esteemed journals, thereby enhancing the quality and rigour of scholarly publications. Overall, his career trajectory exemplifies a steadfast dedication to advancing knowledge, fostering innovation, and making meaningful contributions to the fields of Operations and Supply Chain Management, sustainability science, and beyond. Through his interdisciplinary approach and collaborative endeavours, he continues to shape the landscape of academia and inspire future generations of scholars and practitioners alike.

Jose Arturo Garza-Reyes is a professor of Operations Management and head of the Centre for Supply Chain Improvement at the University of Derby, UK. He is actively involved in industrial projects where he combines his knowledge, expertise, and industrial experience in operations management to help organisations achieve excellence in their internal functions and supply chains. He has also led and managed international research projects funded by the British Academy, British Council, European Commission, Innovate UK, and Mexico's National Council of Science and Technology (CONACYT). As a leading academic, he has published over 300 articles in leading scientific journals and international conferences and eight books. He is the associate editor of the *International Journal of Operations and Production Management* and the *Journal of Manufacturing Technology Management*, editor of the *International Journal of Supply Chain and Operations Resilience*, and editor-in-chief of the *International Journal of Industrial Engineering and Operations Management*. His areas of expertise and interest include Operations and Production Management, Supply Chain and Logistics Management, Lean and Agile Operations and Supply Chains, Sustainability within the context of Operations and Supply Chains, Circular or Closed-Loop Operations and Supply Chains, Sustainable and Green Manufacturing, Industry 4.0 Technologies Application in Operations and Supply Chains, Lean Management, Quality Management and Operations Excellence, and Innovation Management.

Acknowledgements

This book would not have been possible without the support and contributions of many individuals, students, researchers, and practitioners. First and foremost, we extend our heartfelt gratitude to our families for their unwavering support, patience, and encouragement throughout this journey.

We are deeply appreciative of our colleagues and collaborators, whose insights and expertise have significantly enriched this work. Special thanks to the authors for their contributions and to the reviewers for their quality feedback during the review process. Without their support, this book would not have become a reality.

We express our gratitude to our publisher, CRC Press of the Taylor & Francis Group, for believing in this project and providing us with the platform to share our work with a wider audience.

To all the readers who embark on this journey with us, we hope this book serves as a valuable resource and inspires further exploration and innovation in the blockchain technology field.

Contributors

Shalini Ajayan
Institutional Effectiveness
  Department
Rabdan Academy
Mohammed Bin Rashid School of
  Government
Dubai, UAE

Ahmed Almaazmi
Derby Business School
University of Derby
Derby, United Kingdom

Witold Bahr
Marketing, Operations and Digital
  Group
Keele Business School
Keele University
Keele, United Kingdom

Sreejith Balasubramanian
Business School
Middlesex University Dubai
Dubai, UAE

George Bennett
Department of Mining and Mineral
  Processing Engineering
University of Dodoma
Dodoma, Tanzania

Jay Daniel
Centre for Supply Chain
  Improvement
College of Business, Law and Social
  Sciences
University of Derby
Derby, United Kingdom

Pankaj Kumar Detwal
Department of Management
  Studies
Indian Institute of Technology
  Roorkee
Uttarakhand, India

Andrew Hirst
Marketing and Operations
  Department
Derby Business School
University of Derby
Derby, United Kingdom

Fredrick Ishengoma
Department of Information System
College of Informatics and Virtual
  Education
University of Dodoma
Dodoma, United Republic of
  Tanzania

Ian McEwan
Azarc Limited (azarc.io)
Cambridge, United Kingdom

Biswaksen Mishra
Operations Management
Management Development Institute
Gurgaon, India

Utkarsh Mittal
Manager of Machine Learning and
 Automation
Gap Inc
New York City, New York

Sushil Mohan
School of Business and Law
University of Brighton
Brighton, United Kingdom

M. Muthukumar
ICAR-National Research Centre
 on Meat
Hyderabad, India

Rakesh G. Nair
Department of Management
 Studies
Indian Institute of Technology
 Roorkee
Uttarakhand, India

Shubhangini Rajput
Operations Management &
 Decision Sciences
FORE School of Management
New Delhi, India

Fredrick Betuel Sawe
Derby Business School
College of Business, Law and Social
 Sciences
University of Derby
Derby, United Kingdom

Martha Shaka
Centre for Research Training in
 Artificial Intelligence
School of Computer Science and
 Information Technology
University College Cork
Cork, Ireland

Deo Shao
Johannesburg Business School
University of Johannesburg
Johannesburg, South Africa

Vinaya Shukla
Faculty of Business and Law
Department of Business and
 Management
Middlesex University
London, United Kingdom

Rajesh Kumar Singh
Operations Management
Management Development Institute
Gurgaon, India

Rageshree Sinha
School of Business and Law
University of Brighton
Brighton, United Kingdom

Sony Sreejith
Business School
Middlesex University Dubai
Dubai, UAE

Andrew Stapleton
Mechanical and Industrial
 Engineering Department
University of Wisconsin La Crosse
La Crosse, Wisconsin

Arvind Upadhyay
Operations, Logistics and Supply Chain Management
Guildhall School of Business and Law
London Metropolitan University
London, United Kingdom

Christian Michael Veasey
Marketing and Operations Department
Derby Business School
University of Derby
Derby, United Kingdom

Amit Kumar Yadav
Jindal Global Business School
OP Jindal Global University
Sonipat, Haryana, India

Sachin Yadav
Operations Management and Supply Chain
Jindal Global Business School
OP Jindal Global University
Sonipat, Haryana, India

Lee Ee Yern
School of Management
College of Business and Law
Coventry University
Coventry, United Kingdom

Jehan Zaib
Derby Business School
University of Derby
Derby, United Kingdom

Part 1

Introduction

Chapter 1

Mapping the knowledge domain of blockchain technology studies
A scientometric analysis

Jay Daniel and Andrew Stapleton

1.1 INTRODUCTION

Since the Bitcoin was introduced by Nakamoto (2008), the blockchain concept emerged as the de facto platform technology of the cryptocurrency Bitcoin. The blockchain architecture is increasingly gaining attention from academics, practitioners, and regulatory agencies. Blockchain technology represents a distributed and immutable ledger of transaction records which should be consensus-based (Notheisen et al., 2017). The technology tracks and traces assets or services, this ensures data integrity as well (Lacity, 2018). In fact, practitioners and industries are realizing the potential opportunities that blockchain technology could bring, prompting them to explore its potential advantages and challenges. As blockchain technology initiatives are in the beginning phase, nevertheless, corporate investment in blockchain technology, both service and platform, was predicted to hit $2.9 billion by the end of 2019, up from $2.1 billion in 2018 and $945 million in 2017, respectively, as stated by the International Data Corp. Distribution (Nash, 2019). Also, the recent report by blockchain market indicates that the global blockchain market size was estimated to be about $7.4 billion in 2022 and predicted to generate revenue of over $94 billion by the end of 2027 (Markets, 2023).

While blockchain technology gained prominence with both practitioners and academics delving into potential opportunities and challenges, there has been little investigation carried out scrutinizing the current state of the art of blockchain technology literature and observing its past years inclination. The study aims to close that gap by investigating a scientometric analysis-based review to divulge the game changer and innovative technology and progress of the blockchain technology field. A synopsis of the blockchain technology from 1970 to 2023 is provided from the Web of Science (WOS) database. In addition, to provide a comprehensive review, the characterization of publications, mapping of cooperation network between countries, highly cited publications and authors, most prominent keywords in this domain, and leading countries active in this research field are explored. Ultimately, this research presents a comprehensive cutting-edge literature review and analysis of this emerging disruptive domain.

Bibliometric analysis and scientometric analysis are both quantitative and methodological approaches used to study and evaluate scientific literature, but they focus on slightly different aspects and have distinct emphases. Bibliometrics primarily deals with the quantitative analysis of bibliographic data, such as citations, publication counts, authorship patterns, and journal impact factors. It involves the examination of publication patterns and relationships among different elements within the scholarly literature. Conversely, scientometrics is a broader field that encompasses bibliometrics but extends beyond it. It involves the quantitative study of science and its processes, including the production, dissemination, and utilization of scientific knowledge. Scientometric analysis not only incorporates bibliometric techniques but also considers broader contexts and factors (Donthu et al., 2021; Mejia et al., 2021).

Utilizing a scientometric analysis is an appropriate method for exploring literature reviews and evaluating the evolution of research trends and performance in specific thematic areas, including aspects like journals, authors, and countries. Combining both bibliometric and scientometric analyses aids in creating systematic patterns that highlight prominent areas within the scientific domain. In this study, we have employed a dual methodology – comprising bibliometric and scientometric analyses – to assess research trends and performance within the realm of blockchain technology.

Following Gligor and Holcomb (2012), the purpose of this study is to offer an immense survey regarding the bibliometric analysis of scientific papers relating blockchain technology, which is captured from WOS database, as well as future research prospects, by demonstrating the valuable information and knowledge related to the blockchain technology domain. These include the characterization of publications, mapping of cooperation network between countries, highly cited publications and authors, most prominent keywords in this domain, and main countries involved in this research domain. In addition, the study uncovers intriguing findings on emerging trends and potential areas for further exploration related to blockchain technology. This examination could aid both practitioners and academics in gaining a deeper understanding of the current landscape of blockchain technology and identifying prospects for future research initiatives.

The following sections of this chapter are structured as follows: Section 2 furnishes a synopsis of the literature pertaining to blockchain technology background and characteristics such as immutability and digital signatures. Sections 3 and 4 delve into the material and method and results, respectively. Ultimately, in Sections 5 and 6, where the discussion, conclusions, and potential avenues for future research are presented.

1.2 BLOCKCHAIN TECHNOLOGY BACKGROUND AND LITERATURE

The literature review begins with a discussion on blockchain technology characteristics such as immutability, digital signatures, hash function, encryption, and concatenation. Then, a critical review of different aspects of blockchain technology is discussed including sustainability, data processing, consensus, and mining, and applications are then discussed in more detail.

1.2.1 Blockchain characteristics: immutability, digital signatures, hash function, encryption, and concatenation

The blockchain possesses characteristics that enable the technology to be both disruptive and revolutionary. Blockchain technology has been referred to as the Internet of Transactions (Mainelli and Milne, 2016). A blockchain serves as a distributed ledger, constituting a chronological chain of "blocks". Each block encapsulates a record of valid network activity since the addition of the last block (Bogart and Rice, 2015). Each block represents an encrypted piece of information. In theory, any participant can contribute to the block, and all participants can review it. However, no single participant possesses the authority to unilaterally modify the blockchain. Consequently, a blockchain functions as an exhaustive and unchangeable chronicle of network activities, accessible to all participants or nodes within the distributed network.

Blockchains operate on a decentralized network, meaning that no single entity, whether public or private, has control or governance over the system. According to Bogart and Rice (2016), this decentralization eliminates the necessity for third-party intermediaries, reducing friction in various forms such as costs, risk, information, and control in value exchanges. A crucial aspect of blockchain technology is its ability to enable secure value exchanges over the internet between two or more entities, even if they lack a transactional history, without the involvement of a third party like a bank or government. Traditionally, banks act as trusted intermediaries in securing transactions, accompanied by associated costs and fees. In contrast, the blockchain employs a process called "mining" to validate transactions. Multiple miners must unanimously agree on the validity of a transaction, or it is rejected from the block. Mining ensures the integrity and security of transactional information added to the chain, facilitating trust in the blockchain. Trust within the blockchain is established through the mining process, enabling the entire distributed network to reach consensus on data state and network rules without relying on a centralized governing body. Consequently, proposed system improvements can only occur if all actors within the network agree. This results in enhanced transparency and trust within the network, earning the blockchain the moniker of a "trustless" network. In a trustless system, there is no requirement for a central authority, such as a bank or government, to dictate the rules; instead, rules are determined by the network participants, and miners validate transactions accordingly.

Mining involves the process of appending new data blocks to the existing chain through the validation of each node in the network, referred to as miners. A miner incorporates each fresh block into the blockchain after successfully resolving a cryptographic algorithm, a solution that must gain acceptance from all nodes. The network compensates the miner for validating transactions with a form of digital credit, which serves as the primary incentive for miners to consistently engage in mining activities and authenticate the added data. This reward typically takes the form of financial gains. Miners operate as independent entities within the network, and no single miner has the capability to modify the blockchain or introduce invalid data without detection by the rest of the network. In the event of a miner acting as a "bad actor" attempting a fraudulent transaction, the entry is rejected from the blockchain. However, a record of the miner's attempt is retained, allowing the network to identify the miner as a potential threat. This approach significantly enhances traceability within the system.

A blockchain serves as a distributed ledger or database, capturing transactions executed and shared among participating entities. Immutability characterizes the blockchain, implying that no entity, including governments, managers, or administrators, can alter or change it. Unlike conventional systems operating on a single computer, a blockchain operates on every system with access to it. To better understand blockchains, it is important to understand the main components that make it unique.

The first component is digital signatures, these are similar to conventional signatures except they cannot be easily forged. Digital signatures have three main advantages: (1) Authentication – making sure each party is who they say they are; (2) Integrity – ensuring that nothing was altered during the transaction process; and (3) Non-repudiation – since a digital signature is difficult to forge one cannot go back on a transaction. An example of using digital signatures is as follows: Let's say one wants to send 25 bitcoins to one's friend, the first party will be referred to as "she" and the second as "he" throughout this example. Both parties are first going to need to generate a public key and a private key. These keys are generated through some known algorithm and have a mathematical relationship. Next, she is going to take her public key and send it to her friend. You can think of a public key like a user ID, this key is distributed so the entire network has access to it. The public key is essentially a way for parties to send each other data. After she sends her public key to her friend, she is then going to take the transaction data and send it through a hash algorithm. Throughout this discussion, we will be referencing hashing and encryption, when we say "hashing" we want you to think about taking ingredients and baking a loaf of bread, then think about how difficult it would be to take the finished loaf of bread and go back to the original ingredients – in their rawest form. Essentially hashing is a way to ensure data is unable to be decrypted – like the loaf of bread being regenerated into its basic ingredients – since the only way of knowing what the hash represents is by having the original, untouched input. When we talk about encryption, we want you to think of a "lockbox", you put something in the box and lock it using a key. The only way to get back what you put inside is by using the correct key. Encryption and hashing are often used interchangeably in the extant and contemporary literature but are actually very different concepts.

After the first party sends the transaction data through the algorithm, she is going to receive a digest. She will then take this digest, and using her private key encrypt the digest. You can think of a private key as a "password". Importantly, you don't want anybody to have access to this; otherwise, they will have access to your assets in the network. The result of encrypting the digest is the first party's digital signature. Now that she has her digital signature, she is going to take it along with her transaction data and send it to her friend. Her trading partner is going to receive this and take her public key and attempt to decrypt her digital signature. If he is successful, he will get back the same digest that she had previously created using the hash algorithm. If he is unable to decrypt the signature, he knows that the data was not sent from her. This is known as "Authenticates of parties". Let's say he was successful and was able to decrypt her digital signature. Next, he will take the transaction data and send it through the same hash algorithm she used before. He will then compare the output with the digest from the decrypted signature. If they are the same, he knows that the integrity of the transaction was kept. If they are different, he knows that the transaction was altered in some way. Overall, this is how parties interact over the network using blockchain. Next, we will discuss the network's role.

The transaction between the two parties will not just stay between the two of them, it will be broadcasted to the entire network. The network is made up of nodes, these nodes are sometimes called miners – a common term when discussing bitcoin, one of the first contemporary applications of blockchain technology. The nodes, or miners, need to do a couple of things, the first of which is to validate transactions. When miners are validating transactions, they need to check two factors: (1) Does the person sending own what she is sending? And (2) Does the person sending have enough of this to send to the other party? Miners will take the public key of the sender and check all past transactions that are related to this key. If they see that she indeed owns it and has enough of it to send, then the transaction will be accepted by the network. Though there are many types of mining, what we discuss here is called "Proof of work".

After validating numerous transactions, the miners will take thousands of verified transactions and put them into a block. Note that this block has not yet been added to the blockchain. The miners will then take these transactions and, in pairs, hash them together until they have a single digest value. They will then take this digest value along with the hash of the previous block and send them through a cryptographic hash function. Before we go on, let us take a moment to explain what a cryptographic hash function is. A cryptographic hash function is a mathematical function that takes on an arbitrary length input and gives a set length output (Bitcoin uses the hash function SHA-256, for instance). What makes this function unique is that if one changes even one character in the input the hash, or output, will be completely different. As an illustration, if we use the SHA-256 hashing algorithm on the input "I love pizza", it will yield the subsequent hash value.

[1] "c4d9afc24a0f587b9f9855bc88f7c3aa31f811b1bfa465086cf72761f-faf303c".

As you can see, we added an exclamation point to the end and the entire hash is now different. This is what allows blockchain to be immutable. That is, it is a security check against a hacker attempting to steal assets in the blockchain as that hacker would have to change every hash on every block on the blockchain that existed before the transaction. For example, let's say we represent a hacker. We want to change a past transaction, one in which someone had legitimately sent us 25 bitcoins and we want to hack in and change that to 125 bitcoins. When we change this one character, the hash of that current block will completely change. Blockchain is designed where every block is linked through the previous blocks' hash. So, when the hash of the block a hacker is trying to alter changes the chain is now broken. This sends out a red flag to the network that something is not right.

Let's say we got past the point of altering the block; to make this lie a truth, we would need to relink the blocks by recalculating every hash of every block that comes after the one we altered. To do this we would need more than 50% of the networks' computing power, a daunting task. Ultimately, cryptographic hash functions continuously ensure the security of the blockchain as well as making it immutable. Going back to the main tasks of miners, after they take the hash of the previous block, along with the current block's digest, and send it through the cryptographic hash function they will receive an output. Let's call this output the challenge, which is a random value. The miners will take this and concatenate it with a proof, the proof is just some value generated by special software that was created for mining purposes. The miners will then send this through a cryptographic hash function. What they are trying to get is a value that is equal or less than a specific target value. What this target value has is a specific prefix of 0's. One can think of these 0's as coins. For example, let's say the prefix needs to have 40 0's, what this is asking is "how many times would you need to flip the coin to get 40 heads in a row?" This would take around one trillion attempts. The reason miners do this is to solve the proof of work problem. The proof of work problem ensures that there is some organized way to add blocks to the blockchain. If they did not have to solve a proof of work problem, miners would be adding blocks whenever they wanted. The proof of work problem also helps with collision resistance, as well as improves security of the overall network. After a miner solves the proof of work, that miner tells the entire network that they have solved the proof. The entire network of miners immediately stops what they are doing (i.e., attempting to solve the "proof of work problem") and goes through the same process again. The miner that solved the proof of work problem will receive payment in bitcoins, for instance, along with a portion of the transaction fees.

To reiterate, the main components of blockchains are (1) digital signatures: which ensure that transactions maintain integrity, the parties are authentic, and neither party can repudiate the transaction; and (2) cryptographic hash functions: which allow blockchains to be immutable and secure.

Numerous reviews have explored various aspects of blockchain technology. These investigations predominantly centre around different consensus protocols, the currency dimension and its diverse applications, spanning IoT, healthcare, education, voting systems, and government functions (Christidis & Devetsikiotis, 2016; Guru et al., 2023; Khalilov & Levi, 2018; Meng et al., 2018). The security facet has also garnered attention. Additionally, specific reviews have delved into blockchain-based smart contracts (Hewa et al., 2021) and addressed concerns like attacks and vulnerabilities associated with these contracts (Atzei et al., 2017).

Furthermore, there are some blockchain reviews concentrated on various bibliometric analysis on supply chain, blockchain and energy, smart contracts on the blockchain, banking and finance and blockchain, models characterize for blockchain features, etc. (Ante, 2021; Ante et al., 2021; Moosavi et al., 2021; Rico-Peña et al., 2023). The scientometric analysis is proper approach to investigate literature reviews and assessing research evolution trends and research performance in terms of journals, authors, countries, etc. in specific research themes. For mapping the research domain, combination of bibliometric and scientometric analysis will provide systematic patterns empathizing scientific domain hotspots. In this study, we have adopted both methodologies (bibliometric and scientometric analysis) to evaluate research trends and performance in blockchain technology field.

As discussed earlier, the characteristics and components of blockchain technology have many potential applications and benefits in various domains, such as relationship of blockchain and sustainable supply chain (Saberi et al., 2019), a review of blockchain in energy area (Andoni et al., 2019), blockchain systems and data processing (Dinh et al., 2018), blockchain in biomedical and healthcare (Kuo et al., 2017), mining mechanism and strategy for consensus in blockchain networks (Wang et al., 2019), privacy and security issues of Bitcoin (Conti et al., 2018a), blockchain in government for information sharing (Olnes et al., 2017), a review on IoT and blockchain (Ali et al., 2019), blockchain technology adoption in supply chain (Francisco & Swanson, 2018), a research framework for blockchain (Risius & Spohrer, 2017), finance, and digital identity. However, it also faces many challenges and limitations, such as scalability, security, interoperability, and regulation.

1.3 MATERIAL AND METHOD

The examination of literature through the application of statistical and mathematical models is called bibliometric review. In this context, a bibliometric analysis focusing on blockchain has been conducted using VOSviewer. The aim is to explore highly cited authors, their collaborative efforts, key keywords, and the active countries, along with the collaborative networks between countries in this specific field. This literature review adopts a knowledge mapping perspective to evaluate research publications

Table 1.1 Blockchain technology scientometric and bibliometric analysis

Literature review	Scientometric and bibliometric analysis
Blockchain technology/distributed ledger	Publications characterization
	Keywords analysis and co-occurrence
	Co-occurrence of publication sources
	Mapping of cooperation network between countries

within the realms of information systems/technology, finance, business, and management. It evaluates more particularly, in the literature pertinent to the blockchain technology as illustrated in Table 1.1. Different studies adopted the bibliometric and scientometric analyses, followed by qualitative discussion (Figueroa-Rodríguez et al., 2019; Zou & Vu, 2019) which is applied in this study. In addition, the scientometric analysis is the proper approach to investigate literature reviews and assessing research evolution trends and research performance in terms of journals, authors, countries, etc. in specific research themes. For mapping the research domain, combination of bibliometric and scientometric analysis will provide systematic patterns empathizing scientific domain hotspots (Martinez et al., 2019).

1.3.1 Bibliometric analysis

The bibliometric dataset for this review was carried out in the WOS from Thomson Reuters and is recently kept by Clarivate Analytics. It is a well-known database that provides comprehensive and highly calibre records and citation data for numerous academic disciplines (Cañas-Guerrero et al., 2013). The bibliographic dataset for this literature review has been salvaged in August 2023 from the WOS core collection database as follows:

Topics="blockchain technology" AND "distributed ledger"
Timespan="All year"
Database= WOS core collection
Timespan="All year (1970–2023)"
Language=English

The literature search in the WOS database has resulted in 2562 documents. Each document/publication consists of related data to the authors, title, Journal, source, publication year, countries/territories, institutions, citation, keywords, and more parameters. Figure 1.1 shows the blockchain technology areas based on WOS database. The WOS mentioned that "the areas on the chart are not strictly proportional to the values of each entry".

Figure 1.2 demonstrates the bibliometric and scientometric analysis process flow. The process of data collection (documents) commenced with the establishment of sensible inclusion and exclusion criteria to capture pertinent papers from the WOS database. These criteria were delineated as follows: (1) articles had to be published in English; (2) articles should fall within the timeframe of all years, specifically between 1970 and 2023, to effectively reflect the comprehensive and contemporary landscape of the

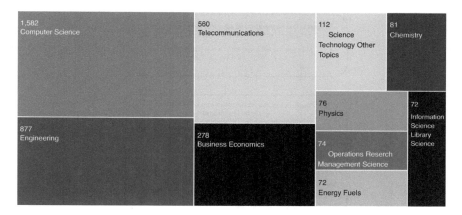

Figure 1.1 Blockchain technology categories.

research domain. If a document was not in English and did not follow the inclusion criteria, it was excluded.

The science mapping was carried out applying VOSViewer. It is a bibliometric and text analysis and mining tool expanded via van Eck and Waltman (2010). It is especially good for comprehensive bibliometric analysis and visualizing large networks. It comprises different features of data visualizing and text mining such as co-citation and co-occurrence analysis. In this literature review, applying VOSViewer different steps were performed as follows:

- The WOS core collection dataset is loaded to the VOSViewer
- Computation, visualizing, and analysis of publication
- Computation, visualizing, and analysis of geographic distribution of publications
- Computation, visualizing, and analysis of highly cited publications and authors cooperation
- Computation, visualizing, and analysis of co-occurrence of keywords, journals, and countries.

Furthermore, a qualitative discussion has been provided after the science mapping and bibliometric analysis to access the ongoing research themes. On top of that the main research domains are identified and further research areas have been proffered.

1.4 RESULTS

Research performance in any research domain can be shown via publication number over the years. In the blockchain technology domain, 2,562 documents were retrieved from the WOS core collection database. These records

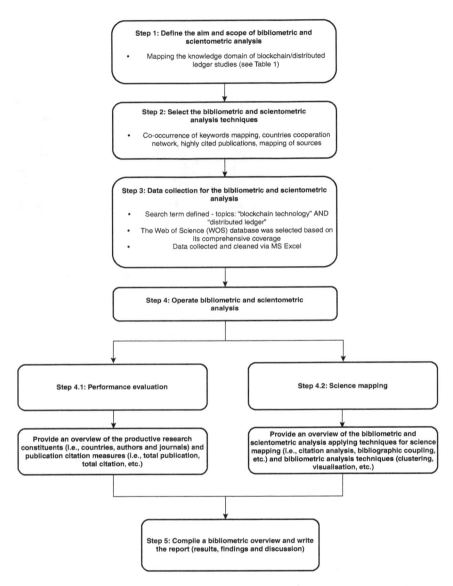

Figure 1.2 Bibliometric and scientometric analysis procedure Source(s): Authors' own work.

show information regarding number of documents and document type published per year in the blockchain technology area as illustrated in Figures 1.3 and 1.4. The yearly trends demonstrate that the first publication emerged in 2014, and since then publications have grown dramatically.

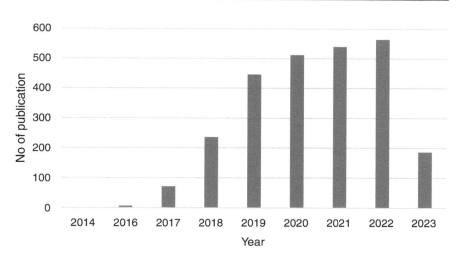

Figure 1.3 Number of publications in the blockchain technology.

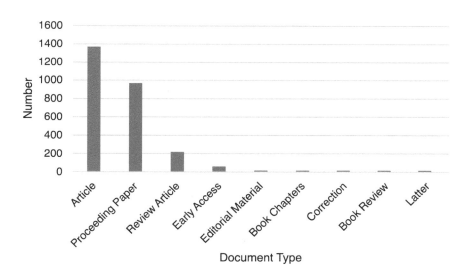

Figure 1.4 Type of document in the blockchain technology.

In order to visualize the relationship between co-authorship and countries, a country-based analysis was carried out to observe co-authorship and country network of publications related to blockchain technology (Figure 1.5). The size of each node displays the contribution to the blockchain technology research domain and thickness of the linkages between regions/countries depicts the cooperation strength between countries.

14 Blockchain Technology

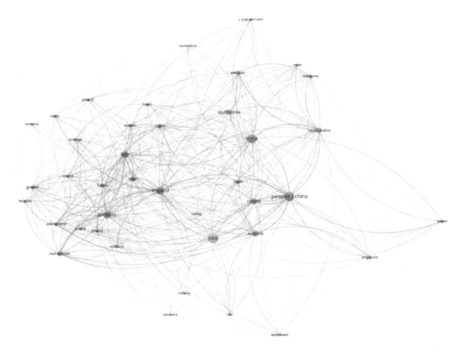

Figure 1.5 Mapping of cooperation network between countries in the blockchain technology domain.

Table 1.2 Countries with high contributions in blockchain technology

Country	Total publication	Cumulative percentage	Total citation	Average article citations	Total link strength
USA	319	19%	8,515	26.7	3,664
China	308	37%	5,277	17.1	2,376
India	258	52%	2,795	10.8	2,257
United Kingdom	185	63%	5,144	27.8	2,181
Germany	138	71%	1,501	10.9	1,162
Italy	136	79%	2,744	20.2	1,521
Korea	88	84%	1,915	21.8	771
Canada	78	89%	1,684	21.6	1,334
Australia	73	93%	1,372	18.8	1,181
Spain	65	97%	854	13.1	714
The Netherlands	51	100%	1,144	22.4	754

The country performance including total publication, citation and average article citation, and total link strength exhibited in Table 1.2. As you can see in Table 1.2, the great influence of the USA, China, UK, and India

Table 1.3 Publication characteristics of blockchain technology field

Publication years	Total publication	Cumulative publication	Percentage of publication	Total author	AU/TP	Total citation	TC/TP
2014	1	1	0.04	3	3	1	1
2016	6	7	0.23	12	2	415	69.17
2017	71	78	2.77	210	2.96	3,384	47.66
2018	236	314	9.21	779	3.30	8,085	34.26
2019	446	760	17.41	1,621	3.63	13,611	30.52
2020	512	1,272	19.98	1,722	3.36	10,069	19.67
2021	540	1,812	21.08	1,968	3.64	5,649	10.46
2022	564	2,376	22.01	2,289	4.06	2,392	4.24
2023	186	2,562	7.26	839	4.51	336	1.81

AU/TP: Authors per published document and TC/TP: Citations per published document.

with about 63% total publication between the 11 countries. Also, these 4 countries have a high linkage between regions/countries which depicts the cooperation strength between countries.

In order to create the map, 20 documents per country were applied so 36 countries were maintained and 5 clusters were emerged. Based on Figure 1.5, it could be concluded that the USA, China, and England have major contribution to the field. Also, other countries such as India, Saudi Arabia, Australia, Canada, Italy, Germany, Pakistan, South Korea, and the Netherlands are the most promising countries in the blockchain technology domain.

A summary of publication characteristics of blockchain technology field is presented in Table 1.3. Just as noted with the overall publication count, the total number of authors witnessed an increase from 2016 to 2022. The average number of authors throughout this timeframe amounted to 1,229, with an average of 3.3 authors per publication. Interestingly, the total citation from 2014 to 2023 is 43,942 and average citation per document is 17.15. Although the number of times cited rising from 2016 to 2019 with highest increase 41,400% in 2016 and lowest increase 68% in 2019. The total citation was declining from 2020 to 2022, respectively, with -26%, -44%, and -58%. Appealingly, the H-index is 96 and g-index is 155 (Egghe, 2006). It can be concluded that there has been a growth trend in blockchain research over these years, by analysing the characteristics of the publication output of blockchain technology.

The 2,562 publications have been authored by 7,106 authors. The top ten highly cited documents have been recognized which is provided in Table 1.4.

The prominent topics displayed within the selected field are represented in Figure 1.9. The most frequently studied keywords are displayed through sizes of nodes and their distances and the lines interconnecting them.

Also, we have visualized co-citation network of cited references/documents in Figure 1.6. The minimum number of cited references is 20 which

Table 1.4 Top ten highly cited publications in the blockchain technology

Title	Authors	Journal/source title	Publication year	Total citations	Average per year
"Blockchain technology and its relationships to sustainable supply chain management" (Saberi et al., 2019)	"Saberi, Sara; Kouhizadeh, Mahtab; Sarkis, Joseph; Shen, Lejia"	"International Journal of Production Research"	2019	1,139	227.8
"Blockchain technology in the energy sector: A systematic review of challenges and opportunities" (Andoni et al., 2019)	"Andoni, Merlinda; Robu, Valentin; Flynn, David; Abram, Simone; Geach, Dale; Jenkins, David; McCallum, Peter; Peacock, Andrew"	"Renewable and Sustainable Energy Reviews"	2019	867	173.4
"Untangling Blockchain: A Data Processing View of Blockchain Systems" (Dinh et al., 2018)	"Tien Tuan Anh Dinh; Liu, Rui; Zhang, Meihui; Chen, Gang; Ooi, Beng Chin; Wang, Ji"	"IEEE Transactions on Knowledge And Data Engineering"	2018	468	78
"Blockchain distributed ledger technologies for biomedical and health care applications" (Kuo et al., 2017)	"Kuo, Tsung-Ting; Kim, Hyeon-Eui; Ohno-Machado, Lucila"	"Journal of the American Medical Informatics Association"	2017	463	66.14
"A Survey on Consensus Mechanisms and Mining Strategy Management in Blockchain Networks" (Wang et al., 2019)	"Wang, Wenbo; Hoang, Dinh Thai; Hu, Peizhao; Xiong, Zehui; Niyato, Dusit; Wang, Ping; Wen, Yonggang; Kim, Dong In"	"IEEE Access"	2019	416	83.2
"A Survey on Security and Privacy Issues of Bitcoin" (Conti et al., 2018b)	"Conti, Mauro; Kumar, E. Sandeep; Lal, Chhagan; Ruj, Sushmita"	"IEEE Communications Surveys and Tutorials"	2018	395	65.83
"Blockchain in government: Benefits and implications of distributed ledger technology (DLT) for information sharing" (Olnes et al., 2017)	"Olnes, Svein; Ubacht, Jolien; Janssen, Marijn"	"Government Information Quarterly"	2017	394	56.29
"Applications of Blockchains in the Internet of Things: A Comprehensive Survey" (Ali et al., 2019)	"Ali, Muhammad Salek; Vecchio, Massimo; Pincheira, Miguel; Dolui, Koustabh; Antonelli, Fabio; Rehmani, Mubashir Husain"	"IEEE Communications Surveys and Tutorials"	2019	357	71.4
"The Supply Chain Has No Clothes: Technology Adoption of Blockchain for Supply Chain Transparency" (Francisco & Swanson, 2018)	"Francisco, Kristoffer; Swanson, David"	"Logistics-BASEL"	2018	349	58.17
"A Blockchain Research Framework What We (don't) Know, Where We Go from Here, and How We Will Get There" (Risius & Spohrer, 2017)	"Risius, Marten; Spohrer, Kai"	"Business & Information Systems Engineering"	2017	346	49.43

Mapping the knowledge domain of blockchain technology studies   17

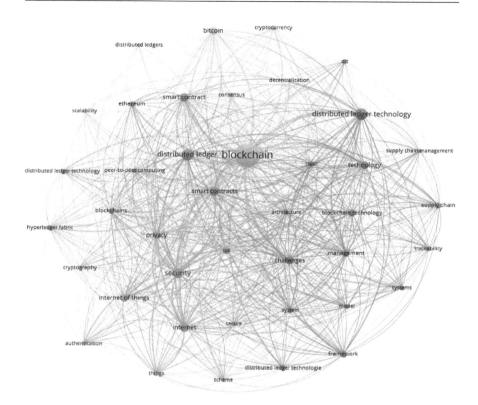

Figure 1.6 Mapping of keywords in the blockchain technology domain.

402 references meet the threshold of the 74,032 references. As you can see in Figure 1.6, the highest co-cited reference/document is creator of bitcoin Nakamoto Satoshi which anonymous creator of bitcoin (Nakamoto, 2008). The second highest co-cited reference/document is "Ethereum: A secure decentralized generalized transaction ledger" (Wood, 2014). The third highest co-cited reference/document is "Hyperledger fabric: a distributed operating system for permissioned blockchains" (Androulaki et al., 2018). The fourth highest co-cited reference/document is "Blockchain: Blueprint for a New Economy" (Swan, 2015), and the fifth highest co-cited reference/document is "Blockchains and smart contracts for the internet of things" (Christidis & Devetsikiotis, 2016).

To examine the interconnections among the keywords used by various authors in the papers, we constructed a Sankey diagram using Bibliometrix which is demonstrated in Figure 1.10. Within the Sankey diagram, the dimensions of the boxes are directly correlated to the frequency of occurrences of each thematic element. The interconnections between the boxes illustrate the progression paths of these themes. Additionally, the thickness of the connecting lines signifies the strength of the relationship between two particular themes (Aria & Cuccurullo, 2017). As depicted in Figure 1.10,

18 Blockchain Technology

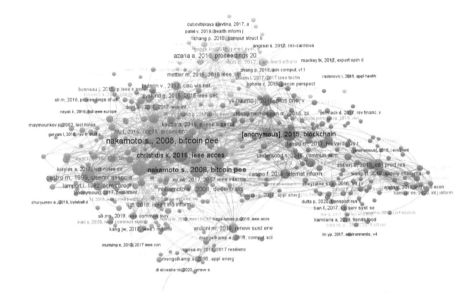

Figure 1.7 Co-citation network of cited references.

the analysis reveals that the keywords "blockchain", "distributed ledger technology", "security", "smart contracts", and "privacy" hold the highest frequency of usage. Notably, these keywords are predominantly associated with the extensively cited works of Chen (2018), Liu et al. (2020), and Treiblmaier (2019).

Moreover, we have illustrated the co-citation network of sources in Figure 1.7 A minimum citation threshold of 20 was applied, and 699 sources met this criterion out of the total 36,698 sources. The top five co-cited sources include IEEE Access, Lecture Notes in Computer Science, IEEE Internet of Things Journal, Future Generation Computer Systems, and IEEE Communications Surveys and Tutorials, respectively.

Furthermore, we have illustrated co-citation network of authors in Figure 1.8 The minimum number of a citation of an author is 20 which 709 authors meet the threshold of the 44,489 authors. The top five highest co-cited authors are Satoshi Nakamoto, Vitalik Buterin, Zibin Zheng, Gavin Wood, and Nir Kshetri, respectively.

1.5 DISCUSSION

In recent years, the blockchain technology is gaining traction from government, academia, and industry (Notheisen et al., 2017). Numerous studies were investigated regarding trust, authorities' corruption, insufficient

Mapping the knowledge domain of blockchain technology studies 19

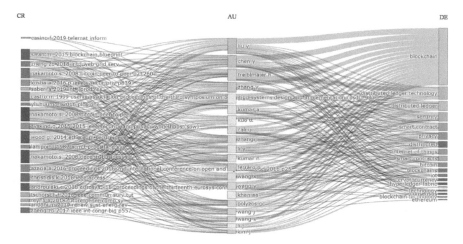

Figure 1.8 Three-field plot: cited references (CR), authors (AU) and keywords (DE).

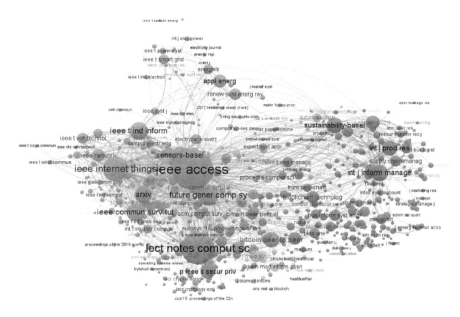

Figure 1.9 Co-citation network of cited sources.

transparency, and issue associating to safety and health of citizen to meet the UN sustainability goals (Hughes et al., 2019; Lacity, 2018; UN, 2019). In the news and mass media, blockchain is repeatedly named a "game changer" (Johnson, 2018). Indeed, the story could be compared to the emergence of the Internet. Both technologies are transformative,

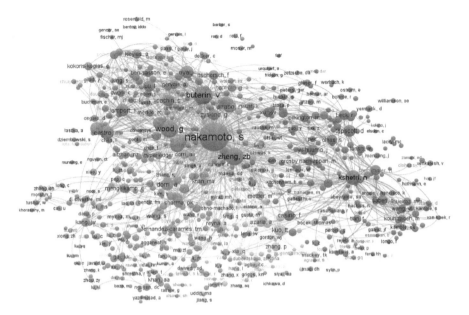

Figure 1.10 Co-citation network of cited authors.

still foundational in essence and are disruptive. There is a rising attention and interest towards the blockchain technology as evidenced by the rapid growth in the publication numbers in this domain from 1 in 2014 to 564 in 2022. In general, articles are rich sources of information and the preferred means of communicating information system research advances.

As you can see in Table 1.5 which summarizes the selected keywords and their quantitative measurements with at least 50 occurrences. Table 1.5 displays the different indices such as occurrence and link strength. The relationship strength between different keywords represents by the total link strength. If the total link strength value is higher, there is a stronger relationship among the keywords (Eck & Waltman, 2018).

In this research, blockchain, distributed ledger(s), distributed ledger technology (DLT), security, internet, smart contract(s), challenges, privacy, management, IOT, bitcoin, framework, and supply chain management have the greatest occurrence and total link strength values. This shows these terms have the most keywords that are interrelated and have the greatest frequency degree in the literature analysed.

In addition, the top cited publications are documents on blockchain for biomedical and healthcare applications, and blockchain in smart energy grids and blockchain in government for information sharing which contextualize keywords such as blockchain technology, DLT, smart contracts, security, and trust. In view of this research findings, the preliminary literature search was limited to blockchain technology and distributed ledger.

Table 1.5 Keywords co-occurrence in the blockchain technology

Keywords	Occurrence	Total link strength
blockchain	1,894	4,440
distributed ledger(s), distributed ledger technology (DLT)	441	1,471
security	335	1,445
internet	268	1,254
smart contract(s)	284	1,104
challenges	218	1,012
privacy	193	817
technology	204	815
management	169	788
internet of things (iot)	185	752
bitcoin	193	683
framework	147	659
supply chain	116	436
Ethereum	117	432
system(s)	90	412
Peer to peer computing	72	395
architecture	71	381
trust	87	356
model	73	326
consensus	93	302
traceability	72	297
cryptocurrency	92	293
scheme	54	254
Hyperledger fabric	87	252
authentication	58	236
scalability	56	221
secure	51	211
decentralization	50	167

These results and findings are lined up with the keywords co-occurrence in the blockchain technology blockchain literature (see Table 1.5).

Keywords have a vital role in defining and reflecting contents of a research. Thus, the analysis of the most pertinent keywords has been applied to determine the skeleton of the important research areas in blockchain technology studies. The hot topics and emerging trends recognized in this research are blockchain and security, smart contracts, challenges, privacy, management, IoT, bitcoin, framework, and supply chain management, model, system, framework, and traceability.

In addition, as you can see in Figure 1.11 which visually displays the prominent sources where the blockchain and ledger technology documents were published with minimum of 10 documents of a source.

22 Blockchain Technology

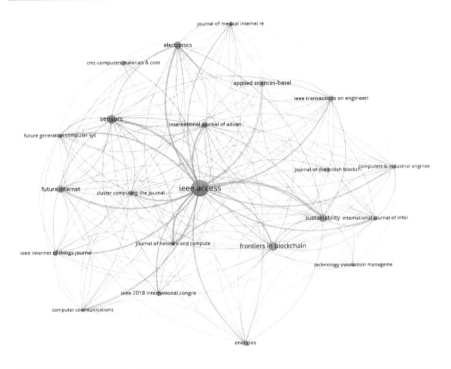

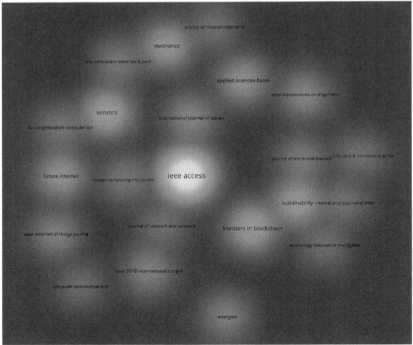

Figure 1.11 Mapping of sources in the Blockchain Technology domain.

There are five clusters, namely, Cluster 1 comprises *Cluster Computing-The Journal of Networks Software Tools and Applications, CMC-Computers Materials & Continua, Electronics, Future Generation Computer Systems, IEEE 2018 International Congress on Cybermatics, EEE Access, International Journal of Advanced Computer Science and Applications, Journal of Medical Internet Research,* and *Journal of Network and Computer Applications and Sensors.* Cluster 2 includes *Computers & Industrial Engineering, Frontiers in Blockchain, IEEE Transactions on Engineering Management, International Journal of Information Management, Journal of The British Blockchain Association, Sustainability,* and *Technology Innovation Management Review.* Cluster 3 has comprised *Computer Communications, Future Internet,* and *IEEE Internet of Things Journal. Cluster 4 has included Applied Sciences-BASEL.* Finally, the cluster 5 included Energies.

Moreover, as you can see in Figure 1.11, IEEE access has the greatest link strength, and it is close-related to its cluster, for example, *Cluster Computing-The Journal of Networks Software Tools and Applications* and *International Journal of Advanced Computer Science and Applications.* However, journals from different clusters could also represent a high level of correlation, such as *Sustainability, Frontiers in Blockchain,* and *IEEE Transactions on Engineering Management.*

Besides, the parameters including number of articles, total citations and total link strength provide insights into the level of productivity.

1.5.1 Managerial and theoretical implications

Our analysis provides managerial insights for industry as well as theoretical direction for the academy. First, managers can improve their processes, especially supply chain, sourcing, and location decision-making by actively applying the blockchain to their firms and partnerships with suppliers and service providers and realizing its full potential. The blockchain offers industry a decentralized, immutable, and secure platform for conducting their operations, especially internationally. Second, managers can transact securely without the cost and delays of a traditional intermediary (e.g., international bank) and can verify the veracity of transactions via the various blockchain proof protocols, while maintaining a permanent record of the entire transaction – from sourcing to the last mile logistics deliveries. This enables easy and fast tracking when needed, such as in the case of a product recall (e.g., food-borne illness). Lastly, the distributed ledger guards against bad actors as the blockchain is immutable and can easily identify an attempted alteration. Regardless of the proof protocol, the nodes can readily kick out any bad actor.

The theoretical implications for the academy derived from our scientometric and bibliometric analysis are also promising for academics. Though we did not specifically cluster the theories used in these publications, there exists ample evidence that future advancement in the blockchain arena can be grounded in Agency Theory as many studies define the contract (e.g., smart contracts) as a unit of analysis. Similarly, Network theory and Neural Networks are a natural fit for grounding future research endeavours in advancing the knowledge domain of blockchains. Other promising landscapes in theory-development include Transaction Cost Economics, Resource-based View, Knowledge-based View, and Dynamic Capabilities Theory. All of these can help to better explain and prescribe the current and emerging phenomena and practices of blockchain development and management. This chapter furnishes valuable insights for both practitioners and academics, offering a thorough analysis of the existing literature on blockchain technology and anticipated future trends. Furthermore, our analysis equips prospective researchers with crucial information, including the strength of linkages, facilitating the identification of promising research areas, and emerging phenomena of interest.

1.6 CONCLUSION

This literature review and scientometric analysis provide valuable information for academics and practitioners in the field of the blockchain technology. This review, scientometric analysis mapping portrays the "as-is" studies and status quo and research trends in this field. The hot topics and emerging trends recognized in this research are blockchain and security, smart contracts, challenges, privacy, management, IoT, bitcoin, framework, and supply chain management, model, system, framework, and traceability. The dataset for this review has been retrieved from the WOS core collection database with topics blockchain technology and distributed ledger.

The knowledge and scientometric mapping of the blockchain technology and distributed ledger dataset revealed the below findings and conclusions:

- There is an observation showing the increasing trend in publications from 2017 up to now.
- The greatest influence documents are "Blockchain technology and its relationships to sustainable supply chain management", Blockchain technology in the energy sector, A data processing view of blockchain systems, "Blockchain for biomedical and health care applications", Consensus and mining strategy management in blockchain networks, Security and privacy issues of bitcoin, Blockchain in government and information sharing, Applications of blockchains in the Internet of Things, Blockchain for supply chain transparency, and Blockchain research framework.
- Keyword analysis and science mapping spotted blockchain technology and DLT, security, internet, smart contract(s), challenges, privacy, IoT, bitcoin, framework, and supply chain management as the terms with the greatest total link strength and occurrence values.

- Keywords were categorized into five main clusters, such as blockchain and supply chain (e.g., DLT, challenges, trust, framework, model, system, and traceability, etc.), research studies related to the security and authentication (e.g., cryptography, IoT, privacy, etc.), smart contracts (e.g., Ethereum, distributed ledger, Hyperledger, etc.), decentralization and scalability (e.g., bitcoin, blockchain, cryptocurrency, distributed ledger), and DLT.
- The most prevalent document types in blockchain technology include articles, conference papers, and review papers.
- The total number of authors increased from 2016 to 2022. The total average of authors during this period was 1,229 and the average per publication was 3.3. Interestingly, the total citation from 2014 to 2023 is 43,942 and the average citation per document is 17.15. Although the number of times cited rising from 2016 to 2019 with highest increase 41,400% in 2016 and lowest increase 68% in 2019. Appealingly, the H-index is 96 and g-index is 155.
- The top five highest co-cited sources in the blockchain are *IEEE Access*, *Lecture Notes in Computer Science*, *IEEE Internet of Things Journal*, *Future Generation Computer Systems*, and *IEEE Communications Surveys and Tutorials*, respectively.
- Blockchain technology research is most prominent in the fields of computer science, engineering, telecommunications, and business economics.
- The top five countries that have largest publications number in the blockchain technology area are USA, China, UK, India, and Germany.
- The hot topics and emerging trends recognized in the blockchain technology research are blockchain scalability and security, smart contracts, privacy and data protection, interoperability and integration with digital twins, and other technologies, i.e., IoT and regulatory concerns.

This study is based on published literature. Additionally, this review only researched documents which are published in English. Overall, this study is a scientometric analysis-based review intended to close the gap of scientific bibliometric review existing in the extant literature. It presents precious information for practitioners and academics with comprehensive analysis of the current blockchain technology literature and future trends. Additionally, our analysis provides future researchers with critical information such as linkages (and strength thereof) in identifying promising research streams and may aid in identifying emerging phenomenon of interest.

We laid the foundation for our analysis by first covering the basic characteristics of blockchains, including immutability, digital signatures, hashing and hash functions, encryption, and concatenation, amongst others. In so doing, it is our hope that the reader has a greater understanding of this revolutionary, disrupting technology that is gaining in prominence as evidenced both in contemporary business, medical, and computing applications and in the exponential growth of this domain in the literature.

NOTE

1 For example: (we used the following website for our example http://www.xorbin.com/tools/sha256-hash-calculator). We put the following into the function: I love pizza ->c4d9afc24a0f587b9f9855bc88f7c3aa 31f811b1bfa465086cf72761ffaf303c When I change one character in the input the following is the result: I love pizza! -> 9a3ac12ec5e859c316ef9e8c 967208c9a426e4a8128e32b3ecc06f80e582c40f.

REFERENCES

Ali, M. S., Vecchio, M., Pincheira, M., Dolui, K., Antonelli, F., & Rehmani, M. H. (2019). Applications of blockchains in the internet of things: A comprehensive survey [Article]. *IEEE Communications Surveys and Tutorials, 21*(2), 1676–1717. https://doi.org/10.1109/comst.2018.2886932.

Andoni, M., Robu, V., Flynn, D., Abram, S., Geach, D., Jenkins, D., McCallum, P., & Peacock, A. (2019). Blockchain technology in the energy sector: A systematic review of challenges and opportunities [Review]. *Renewable & Sustainable Energy Reviews, 100*, 143–174. https://doi.org/10.1016/j.rser.2018.10.014.

Androulaki, E., Barger, A., Bortnikov, V., Cachin, C., Christidis, K., De Caro, A., Enyeart, D., Ferris, C., Laventman, G., & Manevich, Y. (2018). Hyperledger fabric: a distributed operating system for permissioned blockchains. *Proceedings of the Thirteenth EuroSys Conference*, Porto Portugal.

Ante, L. (2021). Smart contracts on the blockchain – A bibliometric analysis and review. *Telematics and Informatics, 57*, 101519. https://doi.org/https://doi.org/10.1016/j.tele.2020.101519.

Ante, L., Steinmetz, F., & Fiedler, I. (2021). Blockchain and energy: A bibliometric analysis and review. *Renewable and Sustainable Energy Reviews, 137*, 110597. https://doi.org/https://doi.org/10.1016/j.rser.2020.110597.

Aria, M., & Cuccurullo, C. (2017). bibliometrix: An R-tool for comprehensive science mapping analysis. *Journal of informetrics, 11*(4), 959–975.

Atzei, N., Bartoletti, M., & Cimoli, T. (2017). A survey of attacks on ethereum smart contracts (sok). *Principles of Security and Trust: 6th International Conference, POST 2017, Held as Part of the European Joint Conferences on Theory and Practice of Software, ETAPS 2017*, Uppsala, Sweden, April 22–29, 2017, Proceedings 6.

Bogart, S., and K. Rice. (2015). *The Blockchain report: Welcome to the internet of value*. Needham & Company, pp. 1–57.

Bogart, S., and K. Rice. (2016). *The blockchain report: Welcome to the internet of value*. Needham.

Cañas-Guerrero, I., Mazarrón, F. R., Pou-Merina, A., Calleja-Perucho, C., & Díaz-Rubio, G. (2013). Bibliometric analysis of research activity in the "Agronomy" category from the Web of Science, 1997–2011. *European Journal of Agronomy, 50*, 19–28. https://doi.org/https://doi.org/10.1016/j.eja.2013.05.002.

Chen, Y. (2018). Blockchain tokens and the potential democratization of entrepreneurship and innovation [Article]. *Business Horizons, 61*(4), 567–575. https://doi.org/10.1016/j.bushor.2018.03.006.

Christidis, K., & Devetsikiotis, M. (2016). Blockchains and Smart Contracts for the Internet of Things. *IEEE Access*, 4, 2292–2303. https://doi.org/10.1109/ACCESS.2016.2566339.

Conti, M., Kumar, E. S., Lal, C., & Ruj, S. (2018a). A Survey on Security and Privacy Issues of Bitcoin [Article]. *IEEE Communications Surveys and Tutorials*, 20(4), 3416–3452. https://doi.org/10.1109/comst.2018.2842460.

Conti, M., Kumar, E. S., Lal, C., & Ruj, S. (2018b). A Survey on Security and Privacy Issues of Bitcoin. *IEEE Communications Surveys & Tutorials*, 20(4), 3416–3452. https://doi.org/10.1109/COMST.2018.2842460.

Dinh, T. T. A., Liu, R., Zhang, M. H., Chen, G., Ooi, B. C., & Wang, J. (2018). Untangling Blockchain: A Data Processing View of Blockchain Systems [Article]. *IEEE Transactions on Knowledge and Data Engineering*, 30(7), 1366–1385. https://doi.org/10.1109/tkde.2017.2781227.

Donthu, N., Kumar, S., Mukherjee, D., Pandey, N., & Lim, W. M. (2021). How to conduct a bibliometric analysis: An overview and guidelines. *Journal of Business Research*, 133, 285–296.

Eck, N. V., & Waltman, L. (2018). *VOSviewer Manual, version 1.6. 8. CWTS, Leiden Google Scholar*, 27, 1–51.

Egghe, L. (2006). Theory and practise of the g-index. *Scientometrics*, 69(1), 131–152. https://doi.org/10.1007/s11192-006-0144-7.

Figueroa-Rodríguez, K. A., Hernández-Rosas, F., Figueroa-Sandoval, B., Velasco-Velasco, J., & Aguilar Rivera, N. (2019). What has been the focus of sugarcane research? A bibliometric overview. *International Journal of Environmental Research and Public Health*, 16(18), 3326.

Francisco, K., & Swanson, D. (2018). The supply chain has no clothes: Technology adoption of blockchain for supply chain transparency. *Logistics*, 2(1), 2. https://doi.org/10.3390/logistics2010002.

Gligor, David M., & Mary C. Holcomb. (2012), Understanding the role of logistics capabilities in achieving supply chain agility: a systematic literature review. *Supply Chain Management: An International Journal* 17, 4438–453.

Guru, A., Mohanta, B. K., Mohapatra, H., Al-Turjman, F., Altrjman, C., & Yadav, A. (2023). A survey on consensus protocols and attacks on blockchain technology. *Applied Sciences*, 13(4), 2604.

Hewa, T., Ylianttila, M., & Liyanage, M. (2021). Survey on blockchain based smart contracts: Applications, opportunities and challenges. *Journal of Network and Computer Applications*, 177, 102857.

Hughes, L., Dwivedi, Y. K., Misra, S. K., Rana, N. P., Raghavan, V., & Akella, V. (2019). Blockchain research, practice and policy: Applications, benefits, limitations, emerging research themes and research agenda. *International Journal of Information Management*, 49, 114–129. https://doi.org/10.1016/j.ijinfomgt.2019.02.005.

Johnson, S. (2018). Beyond the Bitcoin Bubble. *The New York Times*. https://www.nytimes.com/2018/01/16/magazine/beyond-the-bitcoin-bubble.html.

Khalilov, M. C. K., & Levi, A. (2018). A Survey on Anonymity and Privacy in Bitcoin-Like Digital Cash Systems. *IEEE Communications Surveys & Tutorials*, 20(3), 2543–2585. https://doi.org/10.1109/COMST.2018.2818623.

Kuo, T. T., Kim, H. E., & Ohno-Machado, L. (2017). Blockchain distributed ledger technologies for biomedical and health care applications [Review]. *Journal of the American Medical Informatics Association*, 24(6), 1211–1220. https://doi.org/10.1093/jamia/ocx068.

Lacity, M. C. (2018). Addressing key challenges to making enterprise blockchain applications a reality. *MIS Quarterly Executive*, 17(3), 201–222.

Liu, Y., Lu, Q., Paik, H. Y., Xu, X., Chen, S., & Zhu, L. (2020). Design Pattern as a Service for Blockchain-Based Self-Sovereign Identity. *IEEE Software*, 37(5), 30–36. https://doi.org/10.1109/MS.2020.2992783.

Mainelli, M., and Milne, M. (2016). The impact and potential of blockchain on the securities transaction lifecycle (May 9, 2016). SWIFT Institute Working Paper No. 2015-007, Available at SSRN: https://ssrn.com/abstract=2777404.

Markets, M. a. (2023). *Blockchain market by component*. https://www.marketsandmarkets.com/Market-Reports/blockchain-technology-market-90100890.html.

Martinez, S., Delgado, M. D. M., Martinez Marin, R., & Alvarez, S. (2019). Science mapping on the Environmental Footprint: A scientometric analysis-based review. *Ecological Indicators*, 106, 105543. https://doi.org/https://doi.org/10.1016/j.ecolind.2019.105543.

Mejia, C., Wu, M., Zhang, Y., & Kajikawa, Y. (2021). Exploring topics in bibliometric research through citation networks and semantic analysis. *Frontiers in Research Metrics and Analytics*, 6, 742311.

Meng, W., Tischhauser, E. W., Wang, Q., Wang, Y., & Han, J. (2018). When intrusion detection meets blockchain technology: A review. *IEEE Access*, 6, 10179–10188. https://doi.org/10.1109/ACCESS.2018.2799854.

Moosavi, J., Naeni, L. M., Fathollahi-Fard, A. M., & Fiore, U. (2021). Blockchain in supply chain management: a review, bibliometric, and network analysis. *Environmental Science and Pollution Research*. https://doi.org/10.1007/s11356-021-13094-3.

Nakamoto, S. (2008). *Bitcoin: A peer-to-peer electronic cash system*. Retrieved November from www.cryptovest.co.uk.

Nash, K. S. (2019). Business interest in blockchain picks up while cryptocurrency causes conniptions. *Wall Street Journal*. https://blogs.wsj.com/cio/2018/02/06/business-interest-in-blockchain-picks-up-while-cryptocurrency-causes-conniptions/.

Notheisen, B., Hawlitschek, F., & Weinhardt, C. (2017). Breaking down the blockchain hype–towards a blockchain market engineering approach. In Proceedings of the 25th European Conference on Information Systems (ECIS), Guimarães, Portugal, June 5–10, 2017 (pp. 1062–1080). ISBN 978-989-20-7655-3. Research Papers. http://aisel.aisnet.org/ecis2017_rp/69

Olnes, S., Ubacht, J., & Janssen, M. (2017). Blockchain in government: Benefits and implications of distributed ledger technology for information sharing [Editorial Material]. *Government Information Quarterly*, 34(3), 355–364. https://doi.org/10.1016/j.giq.2017.09.007.

Rico-Peña, J. J., Arguedas-Sanz, R., & López-Martín, C. (2023). Models used to characterise blockchain features. A systematic literature review and bibliometric analysis. *Technovation*, 123, 102711. https://doi.org/https://doi.org/10.1016/j.technovation.2023.102711.

Risius, M., & Spohrer, K. (2017). A blockchain research framework what we (don't) know, where we go from here, and how we will get there [Article; Proceedings Paper]. *Business & Information Systems Engineering, 59*(6), 385–409. https://doi.org/10.1007/s12599-017-0506-0.

Saberi, S., Kouhizadeh, M., Sarkis, J., & Shen, L. J. (2019). Blockchain technology and its relationships to sustainable supply chain management [Article]. *International Journal of Production Research, 57*(7), 2117–2135. https://doi.org/10.1080/00207543.2018.1533261.

Swan, M. (2015). *Blockchain: Blueprint for a new economy*. "O'Reilly Media, Inc.".

Treiblmaier, H. (2019). Combining blockchain technology and the physical internet to achieve triple bottom line sustainability: A comprehensive research agenda for modern logistics and supply chain management. *Logistics, 3*(1), 10. https://www.mdpi.com/2305-6290/3/1/10.

UN. (2019). *Sustainable Development Goals - United Nations Development Programme*. UN. Retrieved 2019 from https://www.undp.org/content/undp/en/home/sustainable-development-goals.html.

van Eck, N. J., & Waltman, L. (2010). Software survey: VOSviewer, a computer program for bibliometric mapping [journal article]. *Scientometrics, 84*(2), 523–538. https://doi.org/10.1007/s11192-009-0146-3.

Wang, W., Hoang, D. T., Hu, P., Xiong, Z., Niyato, D., Wang, P., Wen, Y., & Kim, D. I. (2019). A survey on consensus mechanisms and mining strategy management in blockchain networks. *IEEE Access, 7*, 22328–22370. https://doi.org/10.1109/ACCESS.2019.2896108.

Wood, G. (2014). Ethereum: A secure decentralised generalised transaction ledger. *Ethereum ProjJect Yellow Paper, 151*(2014), 1–32.

Zou, X., & Vu, H. L. (2019). Mapping the knowledge domain of road safety studies: A scientometric analysis. *Accident Analysis & Prevention, 132*, 105243. https://doi.org/10.1016/j.aap.2019.07.019.

Part 2

Rethinking strategy and blockchain readiness in the digital world

Chapter 2

Blockchain technology readiness model for supply chain management

Ahmed Almaazmi and Jay Daniel

2.1 INTRODUCTION

Modern technology keeps transforming the world in many ways to add value in different sectors. The impact of technologies on manufacturing has increased the rate of production and the speed in which businesses operate tremendously. Employees are now more efficient than ever before and what used to take hours to do can now be accomplished in minutes if not seconds. Blockchain technology sparked a revolution in the industry and created a new gate for many applications in financial and non-financial sectors. Blockchain is a promising technology and shows a great future for the next global governmental and nongovernmental, financial, and non-financial transactions, evidence shows that blockchain may revolutionize many fields (Paliwal et al., 2020). Blockchain technology enables a fast process cycle (Jabbar et al., 2021), can facilitate the reduction of transaction costs, and enhance overall performance and communication in the supply chain, contributing to economic growth (Upadhyay et al., 2021). The Blockchain is regarded by some as the most revolutionary technological innovation (Wong et al., 2020) since the dawn of the internet and the foundation of 'Web 3.0', poised to usher in the future of the internet (ICAEW, 2020a). Blockchain technology has the potential to provide a much faster and cheaper alternative to traditional cross-border transaction methods. Indeed, while typical money remittance costs might be as high as 20% of the transfer amount, Blockchain may allow for costs just a fraction of that, as well as guaranteed and real-time transaction processing speeds (Reiff, 2020). Based on a peer-to-peer (P2P) topology, Blockchain is a Distributed Ledger technology (DLT) that allows data to be stored globally on thousands of servers – while letting anyone on the network see everyone else's entries in near real-time. That makes it difficult for one user to gain control of the network (Lucas Mearian, 2018). Blockchain technology is the most recent and revolutionary technology to create distributed digital encrypted transaction records or ledgers between entities such as governments, businesses, and individuals over a safe resilient immutable network. One of the main Blockchain technology core functions is the decentralization of data and transactions from a single location to thousands of nodes and locations globally, ruling out any loss of data due to natural disasters, data monopoly, manipulation, and making it available to the public anywhere. Blockchain is a distributed, decentralized, public ledger (Reiff, 2020).

One of the world's critical industries is supply chain management (SCM) where goods are transferred across the globe from one country to another and from producers to end users. SCM is a system that monitors and controls goods and services from one entity to the other. This process starts from collecting raw materials to manufacturing and producing a final product delivered to the end user. Efficient SCM creates a great competitive advantage for firms to streamline internal and external processes to manage waste, increase customer satisfaction and value, and lead the market (Fernando, 2023). For products where human health and safety are critical, Blockchain technology comes into play and gathers essential data across all stakeholders within the cycle and stores it in a secure distributed network to ensure goods are transferred from one side of the world to the other safely and as per health regulations. With the power of Blockchain technology, it is extremely valuable and important for SCM where data will be transparent, available, cannot be changed or manipulated, and saved in a decentralized network globally that is not controlled by any company, entity, or government.

The objective of this chapter is to highlight Blockchain technology's benefits and its applications for diverse sectors and how it can benefit SCM. This chapter will also discuss the readiness of Blockchain technology deployment for supply chain and the main motivations to use Blockchain as technology for different sectors and specifically for SCM. Moreover, the motivation to accept Blockchain as a modern technology will be assessed using both the Technology Acceptance Model (TAM) and the Technology, Organization, and Environment (TOE) framework to explain and predict the factors which possibly stimulate the adoption of new technology, in this case, Blockchain. Some research papers will be explored to explain how researchers used acceptance models to justify the challenges and barriers to implementing Blockchain for supply chains. Moreover, the concept of technology readiness (TR) will be discussed. This chapter will also address the readiness of Blockchain for supply chains and the predominant challenges of Blockchain technology deployment in supply chains.

2.2 BLOCKCHAIN TECHNOLOGY

Blockchain is an emerging technology in Industry 4.0 which has great potential to address some challenges that many different industries are facing. Blockchain can be described as a secure shared ledger in a decentralized system. Blockchain permits a group of participants to share data collectively

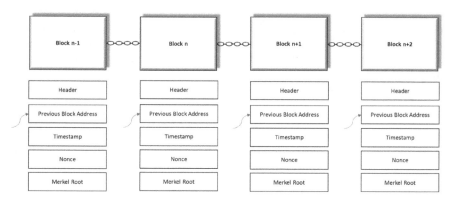

Figure 2.1 Blockchain Structure.
Adopted from Mohitkirange (2022).

in a secure environment. Transactional information from different sources can be collected, modified uniquely, and shared via the Blockchain network. Blockchain enables the sharing of chained recorded information in blocks which are authenticated and validated. Blockchain technology with its core function of decentralization of data and transactions from a central node to thousands of nodes around the globe rules out any loss of data due to natural disasters, data control, manipulation, and makes it available to the public anywhere (Reiff, 2020).

Blockchain data is cryptographically hashed as unique IDs that are divided into shareable blocks that are linked together in a trusted network (Javaid et al., 2021). Because all hashes are created automatically, it is not possible to modify or change any information in the hash. The blocks are chained together in a series and are permanently registered across a peer-to-peer (P2P) network (Steinmetz et al. 2020). Blockchain technology guarantees information integration by creating a single source of authenticating, avoiding data replication, and strengthening security in a peer-to-peer disruption ledger technology structured as a chain of blocks. As a result, Blockchain delivers immutable blocks (Javaid et al., 2021).

2.3 BENEFITS OF BLOCKCHAIN TECHNOLOGY

Blockchain is considered as an innovative and disruptive technology which is going to revolutionize business processes with streamlined efficiency, reliability, and security (Oracle, 2020). Blockchain establishes trust among business partners by offering reliable and shared data and also eliminates data silos in a single system using a distributed ledger shared over a scalable network that only authorized parties can access. Blockchain technology has a high level of data security and generates shared real-time and tamper-evident records among involved participants where data are immutable. Blockchain technology also offers authenticity of records and integrity of information for users and provides faster transaction processing than the initial traditional process from days to minutes, especially between partners for financial settlement (Grecuccio et al., 2020). One of the most valuable benefits of Blockchain for supply chains is its ability to provide seamless traceability and tracking across supply chains. Blockchain can assist in environmental requirements within the supply chain to address hazardous material waste treatment management and manage life cycle assessment (LCA) to achieve operational excellence (Ashraf & Heavey, 2023). In addition, it is used to efficiently manage and allocate energy resources such as grid energy to minimize energy waste and optimize using different energy resources. In the food industry, Blockchain technology can offer traceability and issues of trust to track interactions across supply chain partners associated with commodity details such as the source and ensure that the data uploaded on Blockchain is genuine and recorded with an authenticated entity. Blockchain technology increased transparency in the supply chain industry by its unique capabilities such as traceability and efficiency allowing it to enhance overall food tracking and safety issues and enable better communication and collaboration within the chain (Rosencrance, 2023). Blockchain can significantly change how organizations can run their supply chain transactions with sustainable transparency, traceability, and a secure environment for goods and products as these features are not offered by other technologies. Blockchain technology has great potential to eliminate fraud and deception, solve many existing issues and more due to its benefits and capabilities.

2.4 INDUSTRIAL APPLICATIONS OF BLOCKCHAIN TECHNOLOGY

The industry was not aware of Blockchain's power until it was widely used during the Cryptocurrency boom, such as Bitcoin, which attracted attention from various industries, regarding its potential applications. Blockchain is considered by many as the most revolutionary technological innovation to usher in the future of the internet (ICAEW, 2020b). Blockchain presents a promising technology which can enable many applications in various domains and industries based on its differentiated features. Blockchain offers a unique digital identity which enables authentication for different users or entities such as individuals, organizations, or government entities for land or asset ownership which is protected and encrypted in different Blockchain networks. Blockchain enables a unique feature due to its digital identity and secure authenticated network, known as Smart contact. The smart contract is traceable, secure, and unchangeable due to the features of Blockchain technology where these contracts automate agreement processes between different parties with the contract (Peters & Panayi, 2016). With this said, the applications for Blockchain are extending to different industries; governments and organizations realized these advantages of Blockchain technology and started to explore its benefits in different domains. Blockchain holds the promise of transactional transparency – the ability to create secure, real-time communication networks with partners around the globe to support everything from supply chains to payment networks to real-estate deals and healthcare data sharing (Lucas Mearian, 2018). Some areas where Blockchain applications can be applied are the financial sector, health, and pharmaceutical industry, automotive and manufacturing, government sectors, authenticity and cyber security, supply chain industry (shipping, transportation, logistics), environment, and sustainability.

In the financial sector, Blockchain provides a secure financial transaction between users without referring to a third party such as banks. Blockchain may be used by businesses, organizations, and peer-to-peer global transactions to record, validate, and conduct financial settlements. Blockchain with its success in cryptocurrency has opened the door for digital currency and trading facilitating fast settlement procedures with transparency and authenticity replacing centralized authority.

In the health and pharmaceutical industry, patients' data are considered a very sensitive, private, and crucial aspect of healthcare. Blockchain offers a private network for patient information to overcome the challenges of sharing such sensitive information across healthcare entities without the need for an external third-party entity or systems to export information, hence maintaining privacy. The same also can be applied to pharmaceutical products in the supply chain where challenges arise from altered and incorrect information by some entities within the supply chain cycle to maintain profits and avoid losses due to misuse and handling. With the power of Blockchain technology, the information is distributed and recorded preventing altered information and preserving well-being and trust of pharmaceutical products.

Government sectors are one of the important sectors where Blockchain can offer great value. As discussed earlier, Blockchain technology generates digital identity of ownership and objects that cannot be tampered with in a secure network. Governmental organizations can issue digital identities as discussed earlier for various matters in secure and private networks. The United Arab Emirates (UAE) government, as a part of its digital transformation efforts, decided to capitalize on Blockchain technology to transform government transactions on the federal level, 50% of which will be conducted using Blockchain.

Data security and protection is one of the core functions of Blockchain technology over a shared distributed secure network. This feature differentiates Blockchain technology from the other new innovated technologies. Modifying an existing encrypted record in the chain is impossible based on Blockchain-generated hashes. When new data or transactions are added to the chain, a series of validations will be implemented by various nodes within the Blockchain network to create and approve a link with a previous chain making it secure and inviolate. Data decentralization, security, integrity, transparency, and immutability can be enhanced through blockchain technology (Christopher Lee & Kriscenski, 2019).

Blockchain can revolutionize the supply chain industry with its benefits and capabilities which do not exist in today's traditional supply chain industry, promising major changes in the operation of global technology (Jabbar et al., 2021). Blockchain creates and provides transparency among supply chain networks connecting all parties in the chain cycle (Wong et al., 2020), from original material sourcing to the final service or product for end users, thereby improving trust and authenticity. Moreover, Blockchain enables accurate traceability of products in the whole supply chain cycle (Casino et al., 2020) and increases efficacy and operation addressing various food, pharmaceutical, and agriculture challenges. Blockchain facilitates better communication and collaboration and payment settlement via smart contracts between companies involved, making it faster with lower costs (Al-Jaroodi & Mohamed, 2019). Blockchain digital transactional records offer enhanced security, authenticity, and legitimacy measures which are critical for supply chains (Wong et al., 2020). Blockchain is realized as a resolution for supply chain issues such as traceability problems and generates better and more trustworthy relationships among supply chain partners (Wamba & Queiroz, 2020).

Blockchain technology can serve the environmental and sustainability requirements within the supply chain to address hazardous material waste treatment management and manage LCA to achieve operational excellence. In addition, Blockchain guarantees the traceability of hazardous materials and efficiently manages and allocates energy resources, such as grid energy, to minimize energy waste and optimize the use of different energy resources. Blockchain technology can address various environmental sustainability challenges through applications such as carbon footprint tracking, by providing accurate measurement and record-keeping of carbon emissions data. Also, it enables waste disposal tracking and recycling activities, enhancing waste management and reducing environmental impact. Blockchain can also enhance natural resource management by enabling traceability and sustainability of natural resources.

2.5 BLOCKCHAIN FUTURE DRIVE TECHNOLOGY ADOPTION

Blockchain technology has been a new area of investigation by many researchers across various sectors due to the tremendous value it offers. It addresses numerous challenges in different fields, particularly in the realm of supply chains. The majority of researchers are investigating Blockchain technology in SCM, highlighting its competitive advantages and benefits.

Moreover, the impact of Blockchain on key SCM aspects such as cost, quality, speed, reliability, risks, and sustainability will be examined, along with strategies overcome challenges related to Blockchain adoption in supply chains. Some studies assessed the acceptance of Blockchain technology deployment in SCM.

There are many models that have been developed by researchers to understand and classify variables to identify key factors and barriers to new technology acceptance and adoption. Some of the well-known and famous models which are widely used in assessing new technology acceptance and adoptions in different domains are the TAM by Davis, Bogozzi, and Warshaw in 1989; Unified Theory of Acceptance and Use of Technology (UTAUT) (Venkatesh et al., 2003); Theory of Diffusion of Innovations (DIT) by Rogers in 1995; the Theory of Reasonable Action (TRA) by Fishbein and Ajzen in 1975 (Taherdoost, 2022); and TOE by Tornatsky and Fleischer in 1990 (Tornatsky & Fleischer, 1990).

These models were developed to determine and explain end user's acceptance of using new technology and describe user acceptance behaviour to adopt innovative and new technology. For example, TAM defines the factors into external and internal factors to better understand users' behaviour and define which factors impact users' technology acceptance; key elements include perceived usefulness, ease of use, and attitude (Marangunić & Granić, 2015). UTAUT model, however, tends to look into end user acceptance as performance expectancy, effort expectancy, and social influence to explain user acceptance of new technology adoption (Venkatesh et al., 2003). For example, in a study by Christopher et al. (2019) on the behavioural intention to use Blockchain technology, the UTAUT was employed. The results showed that perceived operational usefulness has a positive influence on blockchain use. Based on their findings, a positive outcome was reached at the end of the use case.

The TOE model was applied in various studies to identify adoption barriers, often in combination with other models or findings from literature reviews by different authors. Kouhizadeh, et al (2021) was used the TOE model alongside force field theory to examine intra-organizational, inter-organizational, system-related, and external barriers, identifying key variables for blockchain adoption in supply chains. These factors included the immaturity of technology, security concerns, management commitment, policies, financial constraints, lack of awareness, government regulations and standardization, ethical practices, and negative perceptions of the technology.

A similar approach also was followed by Farooque et al. (2020) where specific and key barriers were identified to develop a new model explaining the challenges of Blockchain adoption in the supply chain for LCA and how these barriers interact with each other. The authors identified 13 barriers in four main categories intra-organizational, inter-organizational, system-related, and external barriers. The most prominent barriers identified by Farooque et al. (2020) are 'Technology immaturity, lack of government policy and regulation guidance and support, lack of external stakeholder pressure and involvement, and challenges in integrating sustainable practices'.

A study by Kamble et al., (2021) integrated the TOE model with the TAM model to identify the significant factors influencing blockchain acceptance in supply chains. The study combined the factors of the TAM and TOE models, it incorporated factors such as perceived financial benefits, technical know-how, complexity, relative advantage, compatibility, and information security from the technological constructs. From organizational constructs, factors such as training and education as well as top management support were part of the study and finally factors such as competitive pressure and partner readiness from environmental constructs. The outcome of the study identified that partners' readiness, perceived ease of use, competitor pressure, and perceived usefulness are the significant drivers for Blockchain acceptance in the supply chains.

2.6 BLOCKCHAIN READINESS MODEL

TR concept refers to an end user perception of a new technology with two options of either accepting or rejecting the new technology based on the effective values the new technology offers to users. TR measures user satisfaction and reactions to a specific product or service and individual's customer perception. Similarly, to Blockchain as a new technology, TR in SCM will measure personal traits that increase Blockchain adoption to accomplish private or work-related goals (Ruangkanjanases et al., 2022). The readiness of a system or network is the capability to create a process which has flexibility, collaboration, redundancy, and visibility with efficiency that is capable of mitigating new or innovative technology (Tuli, 2023).

The ambition for Blockchain adoption in supply chains can be evaluated in an earlier stage by exploring the readiness of Blockchain technology adoption in supply chains which is derived from the end users and entities within SCM acceptance to deploy Blockchain technology. As highlighted in previous sections, there are many models such as TAM, UTAUT, and TOE which were developed by researchers to address and explain this acceptance behaviour and barriers due to challenges discussed earlier. Different studies were trying to extend these models or create new frameworks to explain Blockchain technology acceptance and readiness for the supply chain.

To understand and facilitate Blockchain adoption in SCM, it becomes essential to investigate supply chain end-to-end cycle readiness before understanding the adoption phase. Some of the major issues and barriers which could impact the readiness of the supply chain are leadership and management support, awareness and knowledge, infrastructure and TR, security and vulnerability, expertise and capabilities, regulatory constraints and policy, cost of implementation, cyber-attacks, environmental sustainability, and collaboration between all entities within the supply chain cycle. Explaining each challenge will help to understand these barriers and have a better picture of its impact on the readiness of Blockchain adoption in the supply chain.

The most appropriate and suitable model which can serve the objective of Blockchain TR for SCM is TOE model by Tornatsky and Fleischer in 1990 (Tornatsky & Fleischer, 1990). Davis (1989) developed the TAM framework and (Venkatesh et al., 2003) developed its extension the UTAUT to understand the factors impacting the need to use a new technology; the other acceptance models were developed consequently. TAM is helpful 'to explain the adoption of a general technology with a theoretical and strong psychometric base and has robust explanatory style' (Chatterjee et al., 2021). TAM also comes with limitations in kerbing external factors and looks at expected future behaviour. Where TOE offers superior attention to both internal and external factors such as environment, organization dimension, and organization strategy and capability (Bryan & Zuva, 2021). Despite the rigorousness of UTAUT model, it has some limitations on theoretical and methodological aspects which were not addressed in further studies. 'UTAUT faced critique with regards to its inability to explain behavioural intention in different settings' (Marikyan & Papagiannidis, 2023). TOE framework can comprehensively describe the variables impacting modern technology adoption decisions in different socio-environmental and technological contexts (Hossain & Quaddus, 2011). TOE framework not only can explain extraneous technological and organizational aspects but also social and environmental aspects (Chatterjee et al., 2021). TOE framework, however, has also some limitations like any other framework and one of them is its general and broad explanation which can be an issue for specific contexts or studies. Despite this limitation, TOE framework fits best to test the readiness of Blockchain technology for supply chain among other developed models.

Accordingly, and reference to the TOE model, the variables will be discussed in detail and segmented into three areas: technology, organization, and environment.

1. **Technology**
    1.1. Technical awareness and knowledge of the technical team and technical manager directly influence leadership and management decisions. Being aware of Blockchain technology and its technical details with its advantages and disadvantages will drastically drive the acceptance and readiness of new technologies by the organization or industry. As the awareness and understanding of Blockchain technology perceived usefulness by technical expertise is high or low, the decision will change accordingly and hence impact the readiness of deployment.
    1.2. Blockchain infrastructure requires high-performance computers with powerful computing capabilities such as processing speed, memory, storage, and environmental requirements such as controlled temperature, cooling, and humidity. The readiness of

such infrastructure will contribute to supply chain Blockchain acceptance and adoption. Compromising these infrastructure requirements with inefficient technological hardware, software, and environmental requirements will lead to undesired and unsuccessful results and outcomes (Sadeghi et al., 2022) and hence the readiness of Blockchain for supply chain.

1.3. One of the most critical and essential aspects of any new or innovative technology is how secure is the system or network and how vulnerable it is to internal and external security and cyber-attacks. With a system like SCM where end-to-end process includes many systems and software to run tasks, record data, and store them either locally or remotely, the safety and confidentiality of this information is going to be vulnerable to modification and storage loss based on today's traditional practice. Understanding the power of Blockchain's security characteristics and technology will stimulate the readiness to adopt Blockchain for supply chain to empower security within the whole chain.

2. **Organization**

    2.1. Leadership and management support in different organizations and industries have a direct impact on the way the organization or industry shapes and navigates in today's industry shift to growth or innovation to explore new horizons. The decision made by leadership will have a direct impact on the company and/or industry to accept and deploy Blockchain SCM or to reject it as a technology.

    2.2. The availability of expertise and capabilities of the resources who will be leading the whole cycle of building a new system starting from planning and development team till operation and maintenance is the core of SCM. Millions of dollars are spent by many organizations in different fields and areas to build expertise and who will be responsible for operating a system to ensure workflow is going smoothly to maintain revenue, company value, and return of investment by shareholders. Without special expertise and capabilities in each specific area within the organization, the operation of the system or company will not meet the objectives and desired outcomes. In technical areas, expertise and capabilities of the workforce are critical and considered a strategic foundation of firm capabilities to lead to superior performance and competitive advantage in the market to add value to consumers and maintain industry leadership. Having adequate and proper resources will have a direct consequence on the readiness of Blockchain adoption in supply chain.

    2.3. The cost of blockchain implementation for SCM also has a direct effect on Blockchain readiness. The cost is considered one of the main elements that many organizations and government entities raise as a concern where the cost in many cases is extremely high where higher management is trying to maintain positive figures for the company's financial performance numbers. The investment is massive and careful decisions need to be taken by higher management to invest the amount which the organization or government leaders see it as a profit or income to operate, upgrade, and maintain positive financial performance and meet key performance indicator which is monitored by stakeholders and ministries. Blockchain technology requires great investments to build the infrastructures operating it and manage skills which will operate the whole system (Lee & Gharehgozli, 2021).

3. **Environment**
   3.1. Regulatory constraints, appropriate regulations, and policies for deploying Blockchain for SCM are extremely vital. As SCM is an end-to-end cyclic chain that touches different entities and countries within its echo system, it must accommodate different requirements and align with multiple associations and organizations. There are various regulations set by governments and the Ministry of Trade to protect national interest and its producers and consumer interest. Moreover, it controls and regulates rules and policies to maintain sustainability and attend to green environmental issues. Since each country has its own regulations and policies which vary from one country to another, these rules and regulations become a constraint for Blockchain readiness and deployment for SCM. Hence, having agreed regulations and policies that are aligned between different governments and institutes nationally and internationally will play an immense role in Blockchain readiness and acceptance for SCM.
   3.2. Environmental sustainability has become a global concern in recent years, raising questions about the impact of manufacturing supply chains on the environment and their sustainability. As a result, it was added as a variable in this model. In a globalized world, environmental sustainability is a critical success factor in different economies and industries. The awareness and the transformation need are increasing to create a more environmentally sustainable society and allow sustainable organizations to thrive to become more sustainable and elevate its standards towards more sustainable manufacturing. One of the Blockchain potential applications to improve supply chain sustainability in the food supply chain is traceability. Blockchain technology can contribute to environmentally sustainable country's development objectives. Blockchain facilitates the realization of a sustainable supply chain by empowering smart cities and improving energy efficient environment. Hence, energy resources such as grid energy become important to minimize energy waste and optimize using different energy resources.

3.3. As supply chain involves many stakeholders, organizations, entities with the process cycle, and different countries, collaboration becomes a key factor to enable the readiness of Blockchain adoption in supply chain. Several types of collaboration among countries, regulatory, global institutes, and entities and partners within the chain need to be at high and extreme level of collaboration to establish positive and constructive relationships to have a significant outcome for Blockchain readiness. The lack of collaboration, communication, and alignment among the supply chain community and its partners will hinder cooperation, thereby affecting the readiness for blockchain deployment.

Figure 2.2 shows the proposed readiness model and the segments of TOE for the selected variables which were discussed above and have an impact on Blockchain readiness for SCM.

As discussed earlier, the above variables are considered as the most critical ones which were selected for the proposed readiness model. Collaboration is closely linked to two other variables: leadership support and regulatory and

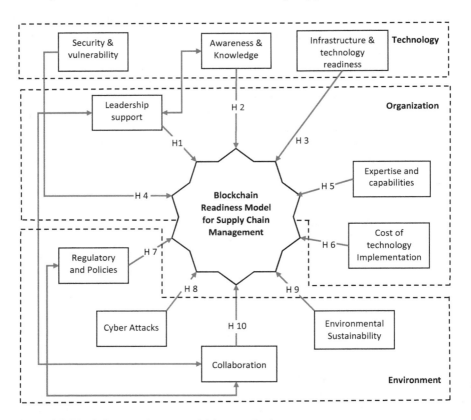

Figure 2.2 Blockchain readiness model for supply chain management.

policies. Without streamlined and coordinated efforts between these key factors, collaboration will not be effective, ultimately hindering the readiness for blockchain adoption in the supply chain. The second correlation among the variables is between leadership support and awareness and knowledge of blockchain technology. Without a clear understanding of the benefits of blockchain to supply chain management (SCM), leadership support will be ineffective, and informed decision-making will not occur. Consequently, the readiness for blockchain in SCM will be delayed or postponed.

2.7 CONCLUSION

Technology transformed our lifestyle and the industry in all aspects and will continue to revolutionize the way we live, work, and or do simple transactions in our daily business or personal tasks and activities. Blockchain technology is one of the innovative technologies that become increasingly popular in recent years where its core function is the decentralization of data and transactions from central nodes to thousands of nodes to overcome any data loss due to natural disasters, data control, and manipulation. The aim of this chapter was to introduce the readiness model for Blockchain adoption in SCM which is a step before major investment by organizations and government entities to assess and decide before moving forward towards adoption. This chapter was designed to explain Blockchain technology and its benefits in different areas specifically in SCM. Moreover, it is important to discuss the benefits that blockchain technology can bring to the supply chain sector, as well as the various applications of blockchain in SCM. Different acceptance models are discussed to explain end user motivation to adopt new technologies. The main variables were discussed in detail explaining their importance and impacts on the proposed readiness model and how to achieve an effective outcome and value from the Blockchain readiness model. The proposed model is a novel contribution and reference for organizations and different governmental and nongovernmental entities to assess its readiness for Blockchain adoption in supply chain.

REFERENCES

Al-Jaroodi, J. and Mohamed, N. (2019) 'Industrial applications of blockchain', *2019 IEEE 9th Annual Computing and Communication Workshop and Conference, CCWC 2019*, pp. 550–555. doi: 10.1109/CCWC.2019.8666530.

Ashraf, M. and Heavey, C. (2023) 'A prototype of supply chain traceability using solana as blockchain and IoT', *Procedia Computer Science*, 217, pp. 948–959. doi: 10.1016/j.procs.2022.12.292.

Bryan, J. D. and Zuva, T. (2021) 'A review on TAM and TOE framework progression and how these models integrate', *Advances in Science, Technology and Engineering Systems Journal*, 6(3), pp. 137–145. doi: 10.25046/aj060316.

Christopher Lee, C., Kriscenski, J. C. and Lim, H. S. (2019) 'An empirical study of behavioral intention to use blockchain technology', *International Business Disciplines*, 14(1), pp. 1–22.

Casino, F. et al. (2020) 'Blockchain-based food supply chain traceability: a case study in the dairy sector', *International Journal of Production Research*, pp. 1–13. doi: 10.1080/00207543.2020.1789238.

Chatterjee, S. et al. (2021) 'Understanding AI adoption in manufacturing and production firms using an integrated TAM-TOE model', *Technological Forecasting and Social Change*, 170(May), p. 120880. doi: 10.1016/j.techfore.2021.120880.

Davis, F. D. (1989) 'Perceived usefulness, perceived ease of use, and user acceptance of information technology', *MIS Quarterly: Management Information Systems*, 13(3), pp. 319–339. doi: 10.2307/249008.

Farooque, M. et al. (2020) 'Fuzzy dematel analysis of barriers to blockchain-based life cycle assessment in China', *Computers and Industrial Engineering*, 147(July), p. 106684. doi: 10.1016/j.cie.2020.106684.

Fernando, J. (2023) *Supply chain management (SCM): How it works and why it is important, investopedia*. Available at: https://www.investopedia.com/terms/s/scm.asp (Accessed: 4 November 2023).

Grecuccio, J. et al. (2020) 'Combining blockchain and iot: Food-chain traceability and beyond', *Energies*, 13(15). doi: 10.3390/en13153820.

Hossain, M. A. and Quaddus, M. (2011) 'The adoption and continued usage intention of RFID: An integrated framework', *Information Technology and People*, 24(3), pp. 236–256. doi: 10.1108/09593841111158365.

ICAEW (2020a) *History of Blockchain, Binance Academy*. Available at: https://www.tradefinanceglobal.com/blockchain/history-of-blockchain/ (Accessed: 27 October 2020).

ICAEW (2020b) *History of blockchain | Technology | ICAEW*. Available at: https://www.icaew.com/technical/technology/blockchain/blockchain-articles/what-is-blockchain/history (Accessed: 16 February 2021).

Jabbar, S. et al. (2021) 'Blockchain-enabled supply chain: analysis, challenges, and future directions', *Multimedia Systems*, 27(4), pp. 787–806. doi: 10.1007/s00530-020-00687-0.

Javaid, M. et al. (2021) 'Blockchain technology applications for Industry 4.0: A literature-based review', *Blockchain: Research and Applications*, 2(4), p. 100027. doi: 10.1016/j.bcra.2021.100027.

Kamble, S. S. et al. (2021) 'A machine learning based approach for predicting blockchain adoption in supply Chain', *Technological Forecasting and Social Change*, 163, p. 120465. doi: 10.1016/j.techfore.2020.120465.

Kouhizadeh, M., Saberi, S. and Sarkis, J. (2021) 'Blockchain technology and the sustainable supply chain: Theoretically exploring adoption barriers', *International Journal of Production Economics*, 231(June 2020), p. 107831. doi: 10.1016/j.ijpe.2020.107831.

Li, K., Lee, J. Y. and Gharehgozli, A. (2021) 'Blockchain in food supply chains: a literature review and synthesis analysis of platforms, benefits and challenges', *International Journal of Production Research*. doi: 10.1080/00207543.2021.1970849.

Lucas, M. (2018) 'What is blockchain? The complete guide | Computerworld'. Available at: https://www.computerworld.com/article/1670066/download-beginners-guide-to-blockchain.html (Accessed: 24 September 2024).

Marangunić, N. and Granić, A. (2015) 'Technology acceptance model: a literature review from 1986 to 2013', *Universal Access in the Information Society*, 14(1), pp. 81–95. doi: 10.1007/s10209-014-0348-1.

Marikyan, D. and Papagiannidis, S. (2023) 'The unified theory of acceptance and use of technology: A review. In S. Papagiannidis' (Ed), *TheoryHub Book*. Available at https://open.ncl.ac.uk / ISBN: 9781739604400, p. 16pg.

Mohitkirange (2022) *Blockchain Structure - GeeksforGeeks*. Available at: https://www.geeksforgeeks.org/blockchain-structure/ (Accessed: 29 November 2023).

Oracle (2020) *What is Blockchain? | Oracle Middle East Regional*, Oracle. Available at: https://www.oracle.com/middleeast/blockchain/what-is-blockchain/#defined (Accessed: 13 August 2023).

Paliwal, V., Chandra, S. and Sharma, S. (2020) 'Blockchain technology for sustainable supply chain management: A systematic literature review and a classification framework', *Sustainability (Switzerland)*, 12(18), pp. 1–39. doi: 10.3390/su12187638.

Peters, G. W. and Panayi, E. (2016) 'Understanding modern banking ledgers through blockchain technologies: Future of transaction processing and smart contracts on the internet of money', *New Economic Windows*, pp. 239–278. doi: 10.1007/978-3-319-42448-4_13.

Reiff, N. (2020) *Forget Bitcoin: Blockchain is the Future*, Investopedia. Available at: https://www.investopedia.com/tech/forget-bitcoin-blockchain-future/ (Accessed: 27 October 2020).

Rosencrance, L. (2023) *7 Applications of Blockchain in the Supply Chain - Techopedia*. Available at: https://www.techopedia.com/7-applications-of-blockchain-in-the-supply-chain (Accessed: 16 July 2023).

Ruangkanjanases, A. et al. (2022) 'Assessing blockchain adoption in supply chain management, antecedent of technology readiness, knowledge sharing and trading need', *Emerging Science Journal*, 6(5), pp. 921–937. doi: 10.28991/ESJ-2022-06-05-01.

Sadeghi, M. et al. (2022) 'Prioritizing requirements for implementing blockchain technology in construction supply chain based on circular economy: Fuzzy ordinal priority approach', *International Journal of Environmental Science and Technology*, 20(5), pp. 4991–5012. doi: 10.1007/s13762-022-04298-2.

Steinmetz, F., Lennart, A. and Ingo, F. (2020) *Blockchain and the Digital Economy: The Socio-Economic Impact of Blockchain Technology*, Aleph. Newcastle Upon Tyne: Agenda Publishing.

Taherdoost, H. (2022) 'A critical review of blockchain acceptance models—blockchain technology adoption frameworks and applications', *Computers*, 11(2). doi: 10.3390/computers11020024.

Tornatsky, L. and Fleischer, M. (1990) *The Processes of Technological Innovation*. Lexington, MA: Lexington Books.

Tuli, P. (2023) *Supply Chain Collaboration in the Era of Blockchain Technologies*, Management for Professionals. Singapore: Springer. doi: 10.1007/978-981-99-0699-4_2.

Upadhyay, A. et al. (2021) 'Blockchain technology and the circular economy: Implications for sustainability and social responsibility', *Journal of Cleaner Production*, 293, p. 126130. doi: 10.1016/j.jclepro.2021.126130.

Venkatesh, V. et al. (2003) 'User acceptance of information technology: Toward a unified view1', *MIS quarterly*, 27(3), pp. 425–478.

Wamba, S. F. and Queiroz, M. M. (2020) 'Blockchain in the operations and supply chain management: Benefits, challenges and future research opportunities', *International Journal of Information Management*, 52. doi: 10.1016/j.ijinfomgt.2019.102064.

Wong, L. W. et al. (2020) 'Time to seize the digital evolution: Adoption of blockchain in operations and supply chain management among Malaysian SMEs', *International Journal of Information Management*, 52(March 2019), p. 101997. doi: 10.1016/j.ijinfomgt.2019.08.005.

Chapter 3

Modeling of blockchain interruptions for smooth integration with supply chain
An ISM-MICMAC mapping approach

Sachin Yadav and Shubhangini Rajput

3.1 INTRODUCTION

In this era, Indian manufacturers are competing in global markets in terms of either the overall cost or quality of the products. Indian manufacturers are trying every possible effort for their sustainability on the international platform. Out of their numerous efforts, Blockchain technology (BCT) is a very famous technology nowadays (Mathivathanan et al., 2021). BC technology is still in its nuance stage (Kouhizadeh et al., 2021; Yadav and Singh, 2019). Manufacturers and stakeholders are very excited to implement BCT in their current running traditional supply chain (TSC) and have a positive attitude toward the new innovative approaches (Francisco & Swanson, 2018). Organizations related to manufacturing, services, etc. are working on how to implement the current trending technology "Blockchain" in their existing traditional and non-digitalized system (Sahebi et al., 2020). Organizations believed that their performance and efficiency would improve drastically after the adoption of BCT. Ivanov et al. (2019) suggested that the role of digitalization and Industry 4.0 will help make a new benchmark for researchers and practitioners. This BCT will give birth to a new paradigm, principles, and models for TSC to reduce risk and ripple effect in the material flow system after integration.

BCT is a technology that stores records in the form of non-volatile (Önder & Treiblmaier, 2018; Yadav & Singh, 2020). It is constructed from millions of blocks and each block has its unique identity, address, and store information in it (Abeyratne & Monfared, 2016). The concept of the cryptographical chain of blocks was introduced by Haber and Stornetta (1992). Later, the BCT's first application is Bitcoin introduced by Nakamoto (2008). The concept of Bitcoin has opened the door of Blockchain for the world. After that Walmart opened its food-based stores to the public by implementing BC with the help of IBM (JOC, 2017). Everledger keeps each diamond's information on a BCT-based supply chain (SC) platform to prevent the occurrence of fraud with customers (Everledger, 2016). Recently TERI, an Indian electricity supply organization has implemented Blockchain with the help of an Indian startup because of its characteristics like transparency in the system, peer-to-peer transactions, static records, elimination of the third party, etc. (Banerjee, 2018; Christidis & Devetsikiotis, 2016; Sahebi et al., 2020). Several detailed studies have been referenced that discuss the influence of blockchain technology (BCT), IoT, Industry 4.0, and other innovations on the TSC, as well as the classification of digital technologies by Andoni et al. (2019), Park and Sung (2020), and Sharma and Joshi (2021); Khan et al. (2023); Xu et al. (2023).

Although this immature innovative trendy technology known as BCT as mentioned above is not in favor of startups, medium- and small-scale firms that are implementing and trying to deploy this technology at a higher cost, consuming more time, and facing a downfall in their performance (Sargent & Breese, 2023). If a startup tries to integrate BCT to improve performance and efficiency, consequently this integrated technology leads to the development of new theories, models, and interruptions in front of SC (Xu et al., 2023) practitioners, startups, and academicians.

This paper does not represent any speculative theory. The objective of this research work is to find out the real-time interruptions while adopting BCT in organizations related to manufacturing, services, etc. in developing countries, i.e., India. In 2018, the Indian Government banned Bitcoin in its Union Budget (Yadav & Singh, 2019). After knowing the demerit of BCT, recently, the Indian Government has recently started to give support to new startups (Yadav & Singh, 2020), and on dated January 21, 2020, the Reserve Bank of India stated that they had not banned any cryptocurrencies, including Bitcoin, in a reply against an appeal filed by the Internet and Mobile Association of India in the Apex court of India in the interest of society. According to the State Ministry for Electronics and IT (MeitY), the Indian government is designing and preparing an approach paper that will discuss the Blockchain framework at the National Level. The potential and importance of distributed ledger technology will be deliberated in the framework of Blockchain and explain the future demand for shared infrastructure for different use cases including targeting the interruption. This clarification is proper signage as motivation for startups in India in the field of BCT. These new startups are doing their work very passionately and grabbing contracts from multinational organizations based on the SC at international platforms (Ante et al., 2018). But due to their excitement and over-commitment about listing themselves in the list of Unicorn and Fortune companies, these startups are ignoring interruptions of BCT that occurred at the time of integration or after integration of this new technology with the existing SC. For taking the title of Unicorn and Fortune, these startups are making minor mistakes that become complex problems later on (Ahluwalia et al, 2020). Finally, these startups are facing various challenges like losing contracts and lack of funds which occurred at that time when these startups failed during the integration of BCT with the TSC.

In the viewpoint of the above, the study provides an understanding of the BCT interruptions that impede BCT adoption in startups. The study also explains the interrelationship of the BCT interruptions while integrating BCT in TSC. Further, ISM-MICMAC approach develops a contextual relationship based structural model and provides solutions to startups that which BCT interruptions require more attention. It will also be fruitful for the startups if the information is sensitive and enables reliable information transfer across the TSC. The study provides insights to startups for BCT interruptions related to speed, visibility of transactions, and traceability. Therefore, the startups built on BCT have the potential to reduce costs and its distributed nature makes the startup systems more resilient, robust, and reliable.

3.1.1 Research gap

BCT is still in its infancy phase (Yadav & Singh, 2022) and only limited work has been carried out. Authors have identified four research gaps after studying multiple papers published in reputed journals in the past and also after discussions with a panel of experts from industries and academics. The following research gap has been identified and enumerated below:

- Yet, limited BCT interruptions have been identified and limited research work has been carried out in this field.
- The only confined qualitative and quantitative approach is applied to Blockchain-based SC.
- No research work has been published based on the interrelationship among interruptions at the micro-level.
- No stress has been put on startups that are working in the field of BCT-integrated TSC.

3.1.2 Research questions

Based on the research gap, the following research questions have been formulated:

- Can the BCT interruptions be examined at the micro-level through startups?
- How TSC will be affected if BCT interruptions are neglected while integrating?
- What are the level of importance and interrelationships among the BCT interruptions?
- Is there potential for overcoming the BCT interruptions to accelerate the BCT implementation in TSC?

3.1.3 Research objectives

The research questions help to derive and raise the following research objectives which are mentioned below:

- To identify crucial interruptions at the micro-level while integrating BCT with TSC through startups.
- To neglect interruption while integrating BCT with TSC is beneficial or not.
- To develop a hierarchy model that shows the interrelationship among interruptions.
- To develop an interrelationship matrix for categorizing the BCT interruptions into dependence, driving, autonomous, and linkage for smooth integration with TSC.

3.1.4 Contributions of the study

The following are the contributions of the study:

- Understanding the BCT interruptions that impede BCT integration in startups for TSC.
- The study explains the BCT interruptions and its contextual relationships based structural model.
- The influential BCT interruptions provide insights to startups for transparency, visibility, and tracking benefits.
- The study classifies the BCT interruptions into strength and dependence through ISM and MICMAC approach.

There are both theoretical and practical implications for startups, policymakers, and decision-makers in prioritizing their efforts for resolving the BCT interruptions while adopting. This study provides the roadmap for effective integration of BCT by identifying the critical BCT interruptions and assessing their interrelationships. Construction of the rest of the research work is as follows: Section 3.2 is constructed based on detailed studies about BCT interruptions that are developed while integrating BCT with TSC. Section 3.3 is designed by authors, in which authors have proposed a methodology for the representation of the interrelationship among various interruptions and proposed a hierarchy model based on the application of ISM and MICMAC. Section 3.4 depicted the result and discussion. The rest of the paper is in Section 3.5 followed by a conclusion, limitations, and future directions for further studies in the future by researchers.

3.2 LITERATURE REVIEW

In the current era, SC companies are facing lots of challenges while adopting new technology known as BCT in their existing TSC mechanisms which are irreversible in nature. For implementing BCT in their TSC (Sharma & Joshi, 2021), organizations publicize the contracts in the open market globally. This is the stage when startups participate in open bids and win the contract. To fulfill the customer demand, viz. transparency in the system

(Khan et al, 2023), static records (Sharma & Joshi, 2021), and high-quality products, these startups (Andoni et al., 2019) start mapping this nuance technology with existing TSC system. Although mapping BCT is a big challenge (Xu et al., 2023) and also a game-changer opportunity for startups (Park & Sung, 2020). In real-time, these startups agreed to sign agreements with any organizations and succeeded in achieving a contract for the implementation of Blockchain (Ahluwalia et al, 2020) in the existing system of the firm. In-ground reality, these startups fail and lots of startups are going to shut down. Since these startups forget the demerits of BCT (Ante et al., 2018; Park & Sung, 2020) and are not ready to face new challenges that occur during integration as well as they have not prepared any strategy in advance to counter the interruptions and have also not prioritized them before countering (Sharma & Joshi, 2021). Significant and critical challenges faced by SC companies are identified and listed in Table 3.1. These interruptions are counted after reviewing numerous published literature, articles, and white papers and concluded after discussion with academic and industry experts having their work experience in the field of the digital supply chain (DSC) with deep knowledge of BCT.

3.3 METHODOLOGY

In this section, first of all, the authors have selected a convenience sample of thirty-two experts, currently working in academics, industry, etc. in the SC field and side by side having knowledge of Blockchain to help in identifying and evaluating the BCT interruptions. The twenty-two industry experts were selected and employed at a tactical level of SCs, have at least 8–10 years of experience, have faced the technology based transitional changes in their respective firms. Besides, eleven academicians from reputed management institutions involved for a minimum period of three years in BCT research projects were selected. But only a total of 25 respondents (18 industry experts and seven academicians) provided us their valuable responses and along with extensive literature review, eight interruptions have been identified.

Post data collection method, authors have used the application of ISM and MICMAC analysis for finding the major and top prior BCT interruptions from the listed eight BCT interruptions mentioned above. ISM is an MCDM method, which was invented by Warfield in 1974 for solving complex problems. ISM model shows the interrelationship among the factors or barriers or interruptions or enablers. Steps of ISM methodology and MICMAC analysis have been taken and followed by Khatwani et al. (2015); Lamba and Singh (2018); and Yadav and Sharma (2017).

Before going into mathematical computation work, a structural self-interaction matrix (SSIM) is prepared after a detailed interview with the expert panel. SSIM is a comparison matrix, constructed with the help of notation, i.e., A, V, X, and O. If element A leads to element B, the sign of notation is V. Element A lags by element B, sign of notation is A. Element A and

Table 3.1 BCT interruptions with their descriptions including references

BCT interruptions and symbols	References	Description
Volatile competitive global market (B 1)	Tapscott and Tapscott (2016); Hileman and Rauchs (2017); Xu et al. (2023)	The nature of market perception is always volatile and the scenario of global demand changes each time. For the integration of Blockchain technology with the supply chain of any organization, the volatile nature of the competitive global market cannot be neglected. The cost of adopting any new technology basically depends upon the competitive global market.
Blockchain architecture (B 2)	Zheng et al. (2017); Cachin, (2016); Sharma and Joshi (2021)	Blockchain architecture is the fundamental design of Blockchain technology. It reflects the structure and complexity of the Blockchain-based supply chain for any organization. Unsolved issues in Blockchain technology created a bottleneck condition at the level of architecture.
Energy consumption (B 3)	Imbault et al. (2017); Truby (2018); Andoni et al. (2019); Sahebi et al. (2020)	Energy consumption puts a direct impact on the cost for any organization, e.g., Bitcoin is an application of BCT consuming electricity more than the demand of the European country "SWITZERLAND" as per a study done by researchers of the University of Cambridge according to the website www.theverge.com. In this case, the energy consumption factor cannot be neglected before the adoption of BCT in the SC of any organization.
Stakeholder Perception (B 4)	Monrat et al. (2019); Sahebi et al. (2020); Sargent and Breese (2023)	The role of Stakeholder perception is very vital for any organization. The involvement of Stakeholders busts up the morale of employees in a firm. Without the involvement or positive perception of stakeholders, it is nearly impossible to introduce BCT into the existing TSC of the firm. Stakeholder plays the role of the key for any supply chain organization adopting new technology.

(Continued)

Table 3.1 (Continued) BCT interruptions with their descriptions including references

BCT interruptions and symbols	References	Description
Non-volatile data (B 5)	Yadav and Singh, (2020); Dorri et al. (2017); Mathivathanan et al. (2021)	Any transaction is stored in the form of a Block. After getting the green signal from the Miners, each Block gets added to the chain and after getting authenticity the Block cannot be deleted or erased, or there is no possibility to make corrections to it. For adding information, each time a new block will be introduced in the chain of blocks.
Blockchain scaling (B 6)	Wright and Filippi (2015); Eyal et al. (2016); Sahebi et al. (2020)	Blockchain Scaling means horizon and vertical scaling of Blockchain technology. Only a limited number of transactions can be processed at any instant in time. It is a big challenge in front of any supply chain organization to fulfill the huge demand of any article at a time, e.g., Bitcoin can process only 4.6 transactions per second according to the website www.hackernoon.com. Keeping in view this picture, millions of articles cannot be processed within a time frame for any BCT-integrated DSC.
BC setup cost (B 7)	Wright and Filippi (2015); Ahram et al. (2017); Xu et al. (2023)	BCT Setup cost means cost from all aspects involved while integrating Blockchain with the TSC of an organization. It covers costs related to architecture, electricity, software, manpower, transaction delay, etc.
Miner's authenticity (B 8)	Iansiti and Lakhani (2017); Nugent et al. (2016); Khan et al. (2023); Yadav and Singh (2023); Yadav and Singh (2024)	Since the block is a non-volatile element in the BCT. The block becomes authenticated after getting the authenticity from Miner. Miners have the authority to validate any transaction in the chain. Authenticity plays a crucial role in this aspect. Fraud may occur at the end of Miner. Due to these reasons, the authenticity of Miner cannot be neglected while adopting BCT with any supply chain system.

element B both are dependent to each other, sign of notation is X. Finally, if element A and element B both are independent, there is no relationship for dependency, represented by O. SSIM matrix is shown in Table 3.2.

In the next stage, an initial reachability matrix is computed with the help of the binary number (0, 1). In this level, V is replaced with 1, A is replaced with 0, X is replaced with 1, and O is replaced with 0. After replacing A, V, X, and O with binary numbers 0 and 1, the authors computed the initial reachability matrix shown in Table 3.3 from the SSIM matrix shown in Table 3.2.

The final reachability matrix is computed by showing the transitivity link from the symbol "*" in Table 3.4. Transitivity is the indirect or hidden dependency relationship among BCT interruptions. If element A leads to element B and element B leads to element C, in this scenario element A must lead to C. For showing this hidden relationship, transitivity is incorporated during computation. The final reachability matrix is shown in Table 3.4.

Table 3.2 Structural self-interaction matrix

	B 1	B 2	B 3	B 4	B 5	B 6	B 7	B 8
B 1	X	A	A	X	A	V	V	O
B 2		X	V	A	V	V	O	O
B 3			X	O	A	V	V	O
B 4				X	O	V	V	V
B 5					X	O	O	A
B 6						X	X	O
B 7							X	X
B 8								X

Table 3.3 Initial reachability matrix

	B 1	B 2	B 3	B 4	B 5	B 6	B 7	B 8
B 1	1	0	0	1	0	1	1	0
B 2	1	1	1	0	1	1	0	0
B 3	1	0	1	0	0	1	1	0
B 4	1	1	0	1	0	1	1	1
B 5	1	0	1	0	1	0	0	0
B 6	0	0	0	0	0	1	1	0
B 7	0	0	0	0	0	1	1	1
B 8	0	0	0	0	1	0	1	1

Table 3.4 Final reachability matrix

	B 1	B 2	B 3	B 4	B 5	B 6	B 7	B 8	Driving power
B 1	1	1*	0	1	0	1	1	1*	6
B 2	1	1	1	1*	1	1	1*	0	7
B 3	0	0	1	1*	0	1	1	1*	5
B 4	1	1	1*	1	1*	1	1	1	8
B 5	1	0	1	1*	1	1*	1*	0	6
B 6	0	0	0	0	0	1	1	1*	3
B 7	0	0	0	0	1*	1	1	1	4
B 8	1*	0	1*	0	1	1*	1	1	6
Dependencies	5	3	5	5	5	8	8	6	45

Transitivity link represented by "*"

Table 3.5 Level partitioning of eight BCT interruptions

Characteristics	Reachability set	Antecedent set	Intersection set	Level
Volatile competitive global market (**B 1**)	1,2,4,8	1,2,4,5,8	1,2,4,8	II
Blockchain architecture (**B 2**)	2,4	2,4	2,4	IV
Energy consumption (**B 3**)	3,4,8	2,3,4,5,8	3,4,8	II
Stakeholder perception (**B 4**)	4	4	4	V
Non-volatile data (**B 5**)	4,5	2,4,5,8	4,5	III
Blockchain scaling (**B 6**)	6,7,8	1,2,3,4,5,6,7,8	6,7,8,	I
BC setup cost (**B 7**)	5,6,7,8	1,2,3,4,5,6,7,8	5,6,7,8,	I
Miner's authenticity (**B 8**)	8	4,8	8	IV

3.4 RESULTS AND DISCUSSION

This section shows the result computed after the adoption of the application of ISM and MICMAC analysis. In Table 3.5, level partition has been carried out based on the reachability set, antecedent set, and intersection set for each BCT interruption. The reachability set for every BCT interruption carries itself and all other elements carry 1 in the corresponding row. The antecedent set for every BCT interruption carries itself and all other elements carry 1 in the corresponding column. The intersection set carries common elements of both the reachability set and the antecedent set. All common elements from the reachability set and intersection set are assigned level one. For the leveling purpose, in the next step, BCT interruptions having level one are eliminated from the whole table for the computation of the next iteration. The same procedures are being followed until the leveling of all interruptions is achieved. In this step of computations, a total of five stages of leveling have been obtained after five iterations. Steps for the iteration have been followed and considered from Chauhan et al. (2019).

The purpose of depicting an ISM model is to obtain the hierarchical level of each BCT interruption so that the interdependencies of each interruption can be analyzed. Such information will assist the industry experts and decision-makers develop appropriate strategies to facilitate the BCT integration in the TSC. The hierarchical model of the ISM illustrating level partitioning of eight BCT interruptions is given in Table 3.5.

The findings reveal five different levels of partitioning for describing the relationship among the BCT interruptions. The first level of the hierarchy includes B 6 and B 7. The second level of the hierarchy includes B 1 and B 3. The third level of the hierarchy includes B 5 and the fourth level of the hierarchy includes B 2 and B 8. The fifth level of the hierarchy includes B 4. From Figure 3.1, it is clearly depicted that B 4 is the top-most driving interruption and is placed at the bottom of the hierarchical model. This infers that stakeholders' perception varies in evaluating the interruptions and could vary between the decision-maker groups, their heterogeneous perspectives and experiences. Even their institutional fields are not completely aligned with the adoption of BCT. Blockchain Architecture (B 2) and Miner's Authenticity (B 8) are also highly driving interruption but less than as compared to Stakeholder Perception (B 4). This illustrates that Blockchain Architecture (B 2) supports traceability, transparency, and also addresses the importance of information sharing in TSC. Secondly, Miner's Authenticity (B 8) verifies that any transaction throughout the SC is authentic or not. It also increases the BCT network security to ensure that no fraudulent activities can take place. Non-volatile data (B 5) is standing at the middle level of

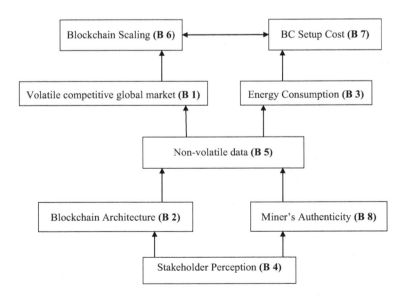

Figure 3.1 Hierarchy of interpretive structural modeling.

the model. BCT interruptions, viz. Blockchain Scaling (B 6) and Blockchain setup cost (B 7) are placed at the top of the hierarchical model.

These BCT interruptions emphasize expanding demands of startups without affecting the performance and ease of TSC. It will also improve the throughput and latency issues in TSC. Besides, complexity involved in deploying the sophisticated technology like BCT involves high level knowledge acquisition and transfer at both managerial and operational levels of TSC. The volatile competitive global market (B 1) and energy consumption (B 3) are just above or at one stage level up as compared to non-volatile data (B 5). This hierarchical model depicts the importance of interruptions among the aforementioned eight interruptions while adopting and implementing BCT in the TSC in any organization in India.

Here, the results of the MICMAC analysis are shown in Figure 3.2. The authors have observed and analyzed that MICMAC analysis is very effective to represent the dependencies and driving power of interruptions in this research work. In this study eight interruptions are divided into four zones, these zones are "autonomous interruption, dependent interruption, linkage interruption, and the fourth zone is driver interruption."

Autonomous zone does not contain any BCT interruptions, indicating that the selected interruptions significantly affect or hinder BCT adoption in TSC. Hence, no BCT interruption is treated isolated in TSC and should be taken into account for the consideration of BCT integration in TSC. B 2 is located in the driving cluster which has high driving power and low dependence suggesting that industry experts and decision-makers should focus on these interruptions and is considered as the most influential BCT interruptions. This is because changes in these interruptions will influence the other BCT interruptions at all levels. B 6 and B 7 are located in the dependency zone which has weak driving power and strong dependence. These BCT interruptions should also be addressed because they have high dependence. Before eliminating them, the BCT interruptions that affect

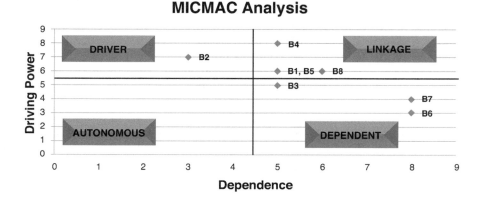

Figure 3.2 MICMAC analysis.

them should be eliminated to increase the chances of successful integration of BCT in TSC. B 1, B 4, B 5, and B 8 are located in the linkage zone, indicating that any changes in these BCT interruptions will affect others at different levels and generate feedback on them.

3.5 MANAGERIAL AND THEORETICAL IMPLICATIONS

This section explains the role of managerial and theoretical contributions in the social environment. Implications will direct the practitioner, researcher, and academician and motivate them to find solutions against the major interruptions. These implications will motivate the academicians and practitioners working in startups to prioritize the solution against interruption so that DSC system will run at an optimized cost. For the development of the right and optimized strategy, stakeholders and the startup implementation team will consider these interruptions in their short-term and long-term plans while integrating BCT with TSC. It is advocated that considering the BCT interruptions will provide startups an insight on what BCT could revamp their TSC and to which influential interruptions the managers must focus on. The findings of the study suggest that the most influential BCT interruptions are currently insufficiently prepared for BCT adoption. This study illustrates that startups are not yet ready to adopt the BCT due to lack of knowledge or expertise and the uncertainties that technology can bring to the startups which would incur the highest cost. Therefore, the industry experts and managers should consider mitigating these interruptions by conducting knowledge transfer sessions about these interruptions. This will enable the startups to understand the new disruptive technology about its functionality and tangible benefits it could bring to the startup firms.

In terms of theoretical contributions, this study explores and explains the significant interruptions to the adoption of the TSC. Secondly, it establishes the contextual relationship among identified BCT interruptions using ISM-MICMAC approach. This approach allows the academicians to classify the identified BCT interruptions into different zones such as autonomous, independent, dependent, and linkage interruptions to understand their behavior. Based on the extensive literature review and data collected from various experts, authors have also computed the driving and dependence power for each BCT interruption in the study and they fell into one of the four zones represented by MICMAC diagram based on their driving and dependence power. Finally, through partitioning of levels for BCT interruptions in the proposed ISM model and their interrelationships across different hierarchies would help researchers to understand the levels and interrelationships between these interruptions. Finally, this research work will motivate and mitigate the discouragement of startups in India or those organizations which are taking back steps or U-turns or shifting their business from the field of BCT due to interruptions of BCT.

3.6 CONCLUSION, LIMITATIONS, AND FUTURE DIRECTIONS

In this manuscript, the authors explain the significant role of interruption in real-time for startups while integrating BCT with TSC. Several detailed studies have been referenced that discuss the influence of blockchain technology (BCT), IoT, Industry 4.0, and other innovations on the TSC, as well as the classification of digital technologies by Andoni et al. (2019), Park and Sung (2020), and Sharma and Joshi (2021); Khan et al. (2023); Xu et al. (2023). BCT has a high potential to redress the complexity in TSC. BCT is attracting stakeholders and startups globally because of its beauty and trying to find out a solution to face global competition and sustain itself on the worldwide platform. There is no doubt that BCT has come in the limelight and integration of this technology is at boom. Although each new invention comes with a few demerits. In this article on account of BCT faults, the authors have discussed the pre- and post-fault of BCT in detail. Practitioners, startups, and BCT implementing teams cannot neglect the interruptions of BCT while implementing BCT in the TSC system. The authors have suggested the qualitative analysis for representing the interrelationship of interruptions and deployed the application of ISM and MICMAC as a solution for transforming TSC into DSC.

In this study, the authors have identified a total of eight BCT interruptions from literature reviews and after discussion with the expert panel. Eight BCT interruptions are as follows: Volatile competitive global market (B 1), Blockchain Architecture (B 2), Energy Consumption (B 3), Stakeholder Perception (B 4), Non-volatile data (B 5), Blockchain Scaling (B 6), BC Setup Cost (B 7), and the eighth BCT interruption is Miner's Authenticity (B 8). After that, ISM and MICMAC analyses are carried out for representing the interrelationship among BCT interruptions. In this study, there are no BCT interruptions found related to the autonomous zone. BCT interruption Blockchain Architecture (B 2) is a major driver of BCT interruption among the eight interruptions. Stakeholder Perception (B 4), Non-volatile data (B 5), **and** Volatile competitive global market (B 1), these three BCT interruptions lie at the intermediate line of linkage and driver zone.

Blockchain Scaling (B 6) **and** BC Setup Cost (B 7), these two BCT interruptions have the nature of dependencies. The energy consumption (B 3) BCT interruption lies at the intermediate line of autonomous and dependence. In this research work, BCT interruption Stakeholder Perception (B 4) cannot be neglected and this interruption is highly driven in nature. Consequently, based on the analysis of interruptions, this research manuscript apparently depicted that BCT is still immature.

3.6.1 Limitations

This research work is based on the sensitive role of interruptions of BCT that develops while integrating it with running TSC where stakeholders want to transform their TSC into DSC. In this research manuscript, qualitative analysis is done with the help of ISM and MICMAC. A few constraints have not been considered, viz. Role of social dimensions, environmental impact, natural climatic disaster, riots, etc. Literature survey, method of investigation, techniques used in the survey analysis, etc. demarcated according to the authors' approach. In this research work, the first and second stage of fuzzy application is not counted while considering the expert's decision through interviews. Experts belong to the field of SC having experience and still working in it and understanding BCT in depth hence experts' knowledge is also a constraint in this research. The strength of experts is 25 in number as this limited number of experts is chosen randomly by authors based on their expertise in the required field.

3.6.2 Future directions

In the future, other MCDM methodologies can also be implemented for the validation of the results and for showing the result comparison. Uncertainty is not counted in this research work hence applications of the Fuzzy System (Fuzzy type I or type II) can be incorporated into further research work extension. In the future, Researchers and Practitioners can use a few new tools like MATLAB, PYTHON, etc. for computational works. Expert panel opinions may vary and can change according to individual knowledge, working field, strengths, and selection methods adopted for the constituted panel therefore panel decisions may differ in the future according to the researcher's approaches. Climatic changes, social issues, and regulation of Govt. bodies are not considered in this research work. In the end, continuous correction is the demand of the global market, and this research will give the direction to do further investigation on the current paradigm. In the end, the developments are still pending and are in the pipeline as per global complexity in the field of the SC, as this BCT 3.0 is in its nuance stage.

REFERENCES

Abeyratne, S. A., & Monfared, R. P. (2016). Blockchain ready manufacturing supply chain using distributed ledger. *International Journal of Research in Engineering and Technology*, 5(9), 1–10.

Ahluwalia, S., Mahto, R. V., & Guerrero, M. (2020). Blockchain technology and startup financing: A transaction cost economics perspective. *Technological Forecasting and Social Change*, 151, 119854.

Ahram, T., Sargolzaei, A., Sargolzaei, S., Daniels, J., & Amaba, B. (2017, June). Blockchain technology innovations. In *2017 IEEE Technology & Engineering Management Conference (TEMSCON)* (pp. 137–141). IEEE. Santa Clara, CA, USA.

Andoni, M., Robu, V., Flynn, D., Abram, S., Geach, D., Jenkins, D., ... & Peacock, A. (2019). Blockchain technology in the energy sector: A systematic review of challenges and opportunities. *Renewable and Sustainable Energy Reviews, 100*, 143–174.

Ante, L., Sandner, P., & Fiedler, I. (2018). Blockchain-based ICOs: Pure hype or the dawn of a new era of startup financing? *Journal of Risk and Financial Management, 11*(4), 80.

Banerjee, A. (2018). Blockchain technology: supply chain insights from ERP. In *Advances in Computers* (Vol. 111, pp. 69–98). Elsevier. https://doi.org/10.1016/bs.adcom.2018.03.007.

Cachin, C. (2016, July). Architecture of the hyperledger blockchain fabric. In *Workshop on Distributed cryptocurrencies and Consensus Ledgers* (Vol. 310, No. 4, pp. 1–4). IBM Research - Zurich CH-8803 Ruschlikon, Switzerland.

Chauhan, A., Kaur, H., Yadav, S., & Jakhar, S. K. (2019). A hybrid model for investigating and selecting a sustainable supply chain for agri-produce in India. *Annals of Operations Research*, 1–22.

Christidis, K., & Devetsikiotis, M. (2016). Blockchains and smart contracts for the internet of things. *IEEE Access, 4*, 2292–2303.

Dorri, A., Kanhere, S. S., Jurdak, R., & Gauravaram, P. (2017, March). Blockchain for IoT security and privacy: The case study of a smart home. In *2017 IEEE International Conference on Pervasive Computing and Communications Workshops (PerCom workshops)* (pp. 618–623). IEEE Kona, HI, USA.

Everledger, Sep. 2016. [Online]. Available: https://everledger.io/

Eyal, I., Gencer, A. E., Sirer, E. G., & Van Renesse, R. (2016). {Bitcoin-NG}: A scalable blockchain protocol. In *13th USENIX Symposium on Networked Systems Design and Implementation (NSDI 16)* (pp. 45–59). Santa Clara, CA.

Francisco, K., & Swanson, D. (2018). The supply chain has no clothes: Technology adoption of blockchain for supply chain transparency. *Logistics, 2*(1), 2.

Haber, S.A., Stornetta Jr, W.S. (1992). U.S. Patent No. 5136646. Washington, DC: U.S. Patent and Trademark Office.

Hileman, G., & Rauchs, M. (2017). *Global Blockchain Benchmarking Study*. Cambridge Centre for Alternative Finance, University of Cambridge, p. 122.

Iansiti, M., & Lakhani, K. R. (2017). The truth about blockchain. *Harvard Business Review, 95*(1), 118–127.

Imbault, F., Swiatek, M., de Beaufort, R., & Plana, R. (2017, June). The green blockchain: Managing decentralized energy production and consumption. In *2017 IEEE International Conference on Environment and Electrical Engineering and 2017 IEEE Industrial and Commercial Power Systems Europe (EEEIC/I&CPS Europe)* (pp. 1–5). IEEE. Milan, Italy.

Ivanov, D., Dolgui, A., Das, A., & Sokolov, B. (2019). Digital supply chain twins: Managing the ripple effect, resilience, and disruption risks by data-driven optimization, simulation, and visibility. In: Dmitry Ivanov, Alexandre Dolgui, Boris Sokolov (Eds.), *Handbook of Ripple Effects in the Supply Chain*, 309–332. Springer.

JOC (2017), "Maersk and IBM team up to digitalize supply chain", *Journal of Commerce*, 18(6), pp. 7–8.

Khan, S., Haleem, A., Husain, Z., Samson, D., & Pathak, R. D. (2023). Barriers to blockchain technology adoption in supply chains: the case of India. *Operations Management Research*, 16(2), 668–683.

Khatwani, G., Singh, S. P., Trivedi, A., & Chauhan, A. (2015). Fuzzy-TISM: A fuzzy extension of TISM for group decision making. *Global Journal of Flexible Systems Management, 16*(1), 97–112.

Kouhizadeh, M., Saberi, S., & Sarkis, J. (2021). Blockchain technology and the sustainable supply chain: Theoretically exploring adoption barriers. *International journal of production economics, 231*, 107831.

Lamba, K., & Singh, S. P. (2018). Modeling big data enablers for operations and supply chain management. *The International Journal of Logistics Management, 29*(2), 629–658.

Mathivathanan, D., Mathiyazhagan, K., Rana, N. P., Khorana, S., & Dwivedi, Y. K. (2021). Barriers to the adoption of blockchain technology in business supply chains: a total interpretive structural modelling (TISM) approach. *International Journal of Production Research, 59*(11), 3338–3359.

Monrat, A. A., Schelén, O., & Andersson, K. (2019). A survey of blockchain from the perspectives of applications, challenges, and opportunities. *IEEE Access, 7*, 117134–117151.

Nakamoto, S. (2008). Bitcoin: A peer-to-peer electronic cash system. *Satoshi Nakamoto*. [ONLINE LINK] https://www.poritz.net/jonathan/past_classes/winter16/CCatRU/BitcoinOriginalPaper.pdf

Nugent, T., Upton, D., & Cimpoesu, M. (2016). Improving data transparency in clinical trials using blockchain smart contracts. *F1000Research, 5*. https://doi.org/10.12688%2Ff1000research.9756.1

Önder, I., & Treiblmaier, H. (2018). Blockchain and tourism: Three research propositions. *Annals of Tourism Research, 72*(C), 180–182.

Park, J. Y., & Sung, C. S. (2020). A business model analysis of blockchain technology-based startup. *Entrepreneurship and Sustainability Issues, 7*(4), 3048–3060.

Sahebi, I. G., Masoomi, B., & Ghorbani, S. (2020). Expert oriented approach for analyzing the blockchain adoption barriers in humanitarian supply chain. *Technology in Society, 63*, 101427.

Sargent, C. S., & Breese, J. L. (2024). Blockchain barriers in supply chain: a literature review. *Journal of Computer Information Systems, 64*(1), 124–135.

Sharma, M., & Joshi, S. (2021). Barriers to blockchain adoption in health-care industry: an Indian perspective. *Journal of Global Operations and Strategic Sourcing, 14*(1), 134–169.

Tapscott, D., & Tapscott, A. (2016). *Blockchain revolution: how the technology behind bitcoin is changing money, business, and the world*. Penguin.

Truby, J. (2018). Decarbonizing Bitcoin: Law and policy choices for reducing the energy consumption of Blockchain technologies and digital currencies. *Energy Research & Social Science, 44*, 399–410.

Wright, A., & De Filippi, P. (2015). Decentralized blockchain technology and the rise of lex cryptographia. *Available at SSRN 2580664*. https://dx.doi.org/10.2139/ssrn.2580664.

Xu, Y., Chong, H. Y., & Chi, M. (2023). Modelling the blockchain adoption barriers in the AEC industry. *Engineering, Construction and Architectural Management, 30*(1), 125–153.

Yadav, S., & Sharma, A. (2017). Modelling of enablers for maintenance management by ISM method. *Ind Eng Manage, 6*(203), 2169–0316.

Yadav, S., & Singh, S. P. (2019). Blockchain critical success factors for sustainable supply chain. *Resources, Conservation and Recycling, 152*, 104505.

Yadav, S., & Singh, S. P. (2020). An integrated fuzzy-ANP and fuzzy-ISM approach using blockchain for sustainable supply chain. Journal of Enterprise Information Management.

Yadav, S., & Singh, S. P. (2022). Modelling procurement problems in the environment of blockchain technology. *Computers & Industrial Engineering, 172*, 108546.

Yadav, S., & Singh, S. P. (2023, December). Importance of Machine Learning for Digital Resilient Supply Chain. In *2023 IEEE International Conference on Industrial Engineering and Engineering Management (IEEM)* (pp. 0001–0005). IEEE Singapore, Singapore.

Yadav, S., & Singh, S. P. (2024). Machine learning-based mathematical model for drugs and equipment resilient supply chain using blockchain. *Annals of Operations Research*, 1–75. https://doi.org/10.1007/s10479-023-05761-0

Zheng, Z., Xie, S., Dai, H., Chen, X., & Wang, H. (2017, June). An overview of blockchain technology: Architecture, consensus, and future trends. In *2017 IEEE International Congress on Big Data (BigData Congress)* (pp. 557–564). IEEE Honolulu, HI, USA.

Chapter 4

Enablers of blockchain adoption in organisations

A view of digital assets and organisational capabilities

Andrew Hirst, Christian Michael Veasey, and Jay Daniel

4.1 INTRODUCTION

The landscape of digital assets is undergoing constant changes and overcoming challenges while nurturing a supportive environment through educational efforts, regulatory measures, and technological advancements. These elements will be crucial in influencing the speed and scope of cryptocurrency adoption (Dourado & Brito, 2014). The term 'cryptocurrency' or 'crypto' refers to a relatively recent form of digital or virtual currency that employs cryptography for security. It is decentralised and operates on a technology called blockchain, which is a distributed ledger enforced by a network of computers (nodes). The most well-known cryptocurrency is Bitcoin, which was introduced in 2009 (Nakamoto, 2008). Today, there are numerous other cryptocurrencies, often referred to as altcoins which have been created, each with its unique features and use cases (Dourado & Brito, 2014). While the concept of cryptocurrency is not entirely new, there is a gap in the literature regarding blockchain adoption within organisations and the understanding of blockchain digital asset enablers which this chapter seeks to address. Thirty prior research papers have identified around eleven types of factors (AlShamsi et al., 2022) that affect the adoption of blockchain. AlShamsi et al. (2022) found that 50% of studies found management involvement to be a critical enabler but recommended that other enablers should be identified to support our understanding of this fast-moving subject area. Our comprehensive search revealed that digital assets were not explicated identified in past research and thus provide a research opportunity and gap in the literature.

In this chapter, we specifically explore both the role of digital assets and management impact on blockchain adoption. Therefore, the contribution of this chapter is two-fold. Firstly, to clarify the role of digital assets, such as cryptocurrencies, in the adoption of blockchain technology. And, secondly to explain in greater detail the role of management in organisational adoption. This chapter is structured as follows. Firstly, we discuss the role of cryptocurrency as a digital asset and form of currency in the contemporary world with an example adopting blockchain in supply chain. Then we examine the methodology and existing frameworks of adoption models for digital assets and blockchain. Finally, we discuss in detail the critical enablers of blockchain technology, and the role digital assets are playing in the development of blockchain and conclude with a summary of critical enablers.

4.2 THE ROLE OF CRYPTOCURRENCY AS A DIGITAL ASSET AND FORM OF CURRENCY IN THE CONTEMPORARY WORLD

The evolution of means of exchanging goods and services has been a continuous process shaped by the needs and complexities of societies. The transition from barter to various forms of money reflects the ingenuity of human societies in facilitating trade and economic growth (Duda, 2015).

Numerous individuals view cryptocurrencies as the new digital money, particularly Bitcoin. Bitcoin is perceived as a constrained and rare asset that is impervious to duplication or replication, making it an efficient 'store of value' (Albrecht et al., 2020). As of 2023, approximately 10,000 distinct cryptocurrencies (Statista, 2023) are contributing to the worldwide advancement of blockchain development. Cryptocurrencies serve several purposes:

1. a store of value in the blockchain, which grows as the network effects increase,
2. a method of recording a digital transaction in the blockchain (a currency of exchange),
3. a speculative investment and trading opportunity.

The multifaceted nature of blockchain has confused legislators, leading to punitive measures against the registration of businesses utilising blockchain currencies. In the US, the US Securities and Exchange Commission (SEC) has issued lawsuits against several platforms, such as Ripple (XRP). The confusion lies in the classification of cryptocurrencies as either a security (or equity stake) or a trading currency (or coin). Currently, Bitcoin (the first cryptocurrency) is named by the SEC as a digital currency. For simplicity, we refer to these cryptocurrencies as digital assets to review the broader nature and purpose of a range of cryptographic applications on the blockchain.

The prominence of digital assets has been notable since the inception of Bitcoin in 2008 (Nakamoto, 2008). During periods of economic uncertainty, digital assets or currencies gained recognition as a store of value and a potential financial safeguard against currency deflation, inflation, and other types of economic instability. Consequently, a variety of other digital currencies have emerged, offering alternative propositions and potential use cases. Digital currency and other cryptographic assets rely on blockchain technology that enables better tracking, accountability, and security (Haynes & Yeoh, 2018). It is argued that this approach enables unsecure systems to operate and support quicker transactions between systems. In part, cryptography is an essential element of a new phase in the World Wide Web, notionally called 'Web3'. Web3 is also associated with the concept of the Metaverse, which professes a seamless virtual world where avatars interact across gaming platforms and other aspects of online social engagement. Hence, there is a prediction that cryptographic assets will emerge as a significant component of our economic development in the future. Nonetheless, cryptographic digital assets (crypto) have been linked to instances of hype bubbles, fraud, and Ponzi schemes. Significant fluctuations in the value of cryptocurrencies have undermined their perceived efficacy as a stable currency. Questions have also arisen regarding their role in the future of the internet. The absence of clear regulations, rules, and definitions further amplifies the risk associated with investment decisions in the crypto space (Haynes & Yeoh, 2018). Even so, Bitcoin enthusiasts, including figures like Michael Saylor, disagree with this concept, they assert that following the abandonment of the gold standard, the US Dollar is prone to devaluation over time, unlike Bitcoin. They argue that successive governments and central banks resort to monetary policy to navigate financial shocks and crises, aiming to sustain liquidity in financial markets and promote lending and investment. This often involves the creation of additional currency to address debt, a practice commonly referred to as quantitative easing.

Quantitative easing tends to benefit borrowers, not savers. Buying assets, such as gold or property, will over the long term increase in value relative to cash if the flow of cash is maintained. Therefore, monetary policy is a key driver of cryptocurrency adoption due to its limited supply as a scarce digital asset. The recent emergence of cryptographic currencies is frequently regarded as the next evolutionary phase, aiming to streamline and enhance the efficiency of transactions. This is particularly pertinent as digital and virtual realms online become progressively ubiquitous facets of human existence.

Cryptographic currencies (crypto) are defined as a decentralised and distributed system that uses cryptographic techniques to create a secure transaction and the exchange of digital tokens (Dourado & Brito, 2014). Bitcoin is acknowledged as the inaugural decentralised cryptocurrency and the source code for Bitcoin is archived on GitHub's open-source code (Nakamoto, 2008). As the code is freely available, other technologists have copied the code and created blockchains with modifications of the Bitcoin code. New cryptocurrencies tend to have unique features and applications (Haynes & Yeoh, 2018). Features such as faster and more energy-efficient transactions are commonly reported. Other developments include the concept of smart contracts and digital tokens such as NFTs (non-fungible tokens) which allow other forms of transactions.

Yilmaz and Hazar (2018) identify five themes driving cryptocurrency adoption: profitability, convenience, anonymity, security, and bookkeeping. However, we also need to consider that crypto currently, due to the absence of clear regulations and price volatility plus its association with money laundering and fraud, has inherent risks (Zhao & Zhang, 2021) that undermine adoption. Despite the risks, many investors see crypto as an opportunity for substantial returns (Sukumaran et al., 2022). Navigating the cryptocurrency investment landscape proves challenging due to the pronounced volatility in both price and volume. However, within various technology sectors, organisations are eager to showcase a sense of FOMO, or 'fear of missing out.'

Initially introduced by Herman (2000), this concept pertains to brand selection and loyalty. FOMO in the context of cryptocurrencies signifies the Fear of Missing Out, depicting the anxiety individuals feel about potentially missing out on profits or exciting opportunities in the crypto market. Also, irrational exuberance provides an overly optimistic and enthusiastic sentiment, often leading to inflated cryptocurrency prices without a solid foundation in fundamentals. These psychological factors driven by social media can contribute to market volatility and sudden price surges, as investors hastily join trends without thorough consideration of underlying values or associated risks (Gupta & Sharma, 2021). Indeed, FOMO has played a role in influencing the adoption of cryptocurrencies in current times (Albrecht et al., 2020). In a traditional network, investments are centralised around the firm's innovation and venture capital allocation. However, in blockchains, the role of venture capital funders is assumed by innovators and investors in the network. These stakeholders are decentralised, extending to individuals or firms seeking success within the network (Alabi, 2017).

Nonetheless, in investor markets, FOMO is commonly known as *irrational exuberance*. Irrational exuberance refers to the unfounded optimism that precedes technology innovation with little or no real foundation (Shiller, 2014; Shiller 2015). Furthermore, Shiller (2023) described Bitcoin as an example of irrational exuberance. Additional research recommends a more comprehensive and balanced understanding of blockchain within the financial ecosystem to understand what drives this adoption other than a store of value (Albrecht et al., 2020). Further factors influencing technology adoption encompass the escalating speed and efficiency of technology, facilitating increased utilisation and sophistication across the phases of collectable, medium of exchange, and ultimately, unit of account.

4.2.1 Facilitating conditions for innovators and early adopters

Employing Rogers' (1983) diffusion concepts, early adopters comprised a varied demographic, including individuals, businesses, investors, and cryptocurrency miners. Innovators and early adopters share a common trait of being risk-tolerant, leading them to be more open to adopting unproven or initial versions of technology and prototypes. Also, innovators or pioneers were driven by the decentralised nature of Bitcoin and the desire to shift away from traditional central banking concepts (Laurence, 2023). The speculative characteristics and the potential for significant returns further captivated interest in cryptocurrencies. Nonetheless, the absence of regulation has posed potential challenges for widespread adoption and for the 'early majority' investors who seek assurance within the established financial system away from illegal activities and fraud (Chuen et al., 2018). Doubters frequently label cryptocurrencies as a Ponzi scheme, while some question the practicality of on-chain coins due to the perceived difficulty in using these assets. Abdeldayem and Aldulaimi (2023) contend that modifying regulations could enhance investor sentiment. In Europe and various other trading regions, there is an ongoing development of new regulations aimed at fostering innovation and adoption. An example is the European Union's regulation called Markets in Crypto Assets (MiCA). MiCA sets out definitions of crypto assets such as e-money (Vicente, 2023), these are:

1. Crypto assets – a digital representation of a value or of a right that can be transferred and stored electronically using distributed ledger technology, or similar technology.
2. Asset-reference tokens – asset-referenced token' means a type of crypto asset that is not an electronic money token and that purports to maintain a stable value by referencing another value, or right, or a combination thereof, including one or more official currencies.
3. e-money – electronic money token' or 'e-money token' means a type of crypto asset that purports to maintain a stable value by referencing the value of one official currency.
4. 'Utility token' means a type of crypto asset that is only intended to provide access to a goods or a service supplied by its issuer.

The development of crypto assets has for many years been a two-sided market development problem. The EU regulation on markets in crypto assets (MiCA) in December 2024 defined crypto assets as digital representations of values or rights, storable and transferrable electronically, using a digital ledger technology. The MiCA indicates that robust regulations are necessary to ensure businesses are compliant with laws and regulations, such as anti-money laundering (AML) and know-your-customer (KYC) regulations (EPRS, 2023).

4.2.2 Technological enablers for adopting blockchain in the supply chain

In an era of global trade, components and finished goods must pass through many ports and processes to reach end uses. A mechanism that records these transitions without causing bottlenecks. Whether a supply chain is ready for blockchain adoption depends on a range of factors, encompassing aspects of technology, organisation, and the surrounding environment. Here we break down the key enablers of the use of blockchain in the supply chain domain.

4.2.2.1 Data infrastructure and process efficiency

Existing systems and their compatibility with blockchain integration are crucial. Smooth data integration and management capabilities are essential. The organisation should have the capacity to collect, store, and manage large volumes of data securely and efficiently on the blockchain. For example, the implementation of Enterprise Resource Planning (ERP) frequently entails the reengineering of business processes, resulting in the establishment of standardised and streamlined workflows. This enhanced efficiency contributes to reduced lead times, minimised wastage, and decreased operational costs (Al-Mashari et al., 2006).

4.2.2.2 Collaboration and partnership

Openness to collaboration with partners using different blockchain systems is essential for wider adoption and increased network value. Exploring integration solutions and pilot projects with potential partners can be beneficial. For instance, systems bring together departments and functions, promoting collaboration and the sharing of information across various functions. This integration optimises coordination within the supply chain, breaking down silos, and improving communication. (Gunasekaran et al., 2004).

4.2.2.3 Technical expertise

Implementing and maintaining a blockchain requires specialised skills and knowledge. Availability of talent within the organisation or through partnerships is key. Ideally, the organisation should have in-house personnel with expertise in blockchain technology, cryptography, and distributed ledger systems. If such expertise is lacking, external consultants or partnerships can be leveraged. Also, depending on the chosen blockchain platform and desired functionalities, custom development skills may be required to build applications and integrate them with existing systems. Furthermore, continuous monitoring and maintenance of the blockchain network and applications are crucial. Adopting blockchain in supply chain management is considered reliant on possessing both technical expertise and accessibility (Wong et al., 2020).

4.2.3 Blockchain adoption in supply chain

As an example of performance expectation, we offer a case study on one aspect of business that shows promising opportunities. Supply chain management encompasses numerous internal and external stakeholders, making blockchain an ideal solution due to its ability to accommodate inputs from diverse participants. Figure 4.1 illustrates a conventional supply chain, where a sequence of interactions takes place sequentially among various stakeholders.

Transactions among these nodes are verified only upon payment through a third-party financial institution in the traditional model. Information flows through various stakeholders, from consumers to retailers, wholesalers/manufacturers, transportation and freight providers, suppliers, and finally resource providers. These transactions traverse multiple systems, and the accuracy of information exchange relies on trust between stakeholders. The lack of traceability poses a risk of misinformation or alteration within the supply chain. The proposed solution advocates for organisations to adopt a private blockchain model. Additionally, it emphasises the need for governance and business standards within a virtual community.

The private blockchain developed by Maroun and Daniel (2019) is presented in Figure 4.2. This is decentralised, connecting all supply chain stakeholders through the blockchain network. Every transaction within the supply chain is stored and verified by all users, ensuring enhanced security

Figure 4.1 Conventional supply chain with transaction verification conducted by a third party, such as a bank. (Maroun & Daniel, 2019).

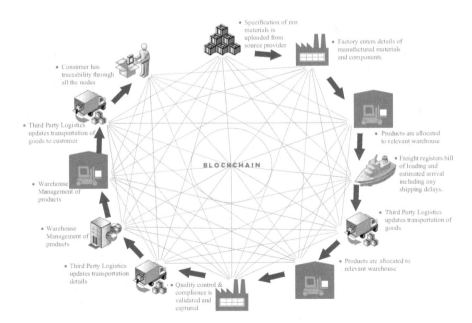

Figure 4.2 Adopting blockchain technology in the supply chain (Maroun & Daniel, 2019).

and transparency. Decentralisation allows each member of the supply chain to access crucial information at any given time. For instance, manufacturers can trace the origin of raw materials, while consumers at the end of the chain can track every step of a product, from raw materials to manufacturing and delivery. As a private blockchain, it facilitates stakeholders exclusively involved in the supply chain to input information such as raw material sources, component manufacturing locations, stock availability, and minimum order quantities. Suppliers of components can include details of quality assurance and testing, and freight forwarders can input information about goods pickup, expected delivery timeframes, and any potential delays.

Operating on a decentralised network without central control, the blockchain ensures secure transactions, and data integrity is maintained through validation by various network stakeholders. Once a record is duplicated on all network servers, altering block content becomes impossible without authorisation from all connected nodes.

As each activity progresses through the supply chain, the private blockchain interconnects all network participants, fostering transparency, validity, and risk reduction among all parties. However, this approach necessitates the active participation of all stakeholders, as it relies on full engagement. Without the inclusion of all relevant parties, such as the freight organisation, gaps may emerge, leading to a lack of complete transparency from start to finish.

The incorporation of blockchain technology enhances the transparency of the supply chain for goods or services. Each block in the chain empowers stakeholders to manage information securely, ensuring an auditable and immutable record. The proposed method comprehensively involves all parties in supply chain management, preventing overlooked transactions, errors, or unauthorised transactions. blockchain's significant contribution lies in guaranteeing the validity and accountability of transactions by the pertinent stakeholders. This technology provides businesses with the capability to oversee the entire supply chain. Table 4.1 shows the enablers of blockchain based on this approach.

The blockchain technology adoption model in the supply chain, as per this approach, is illustrated in Figure 4.2. Subsequently, we will analyse prevailing frameworks of adoption models centred around digital assets and cryptocurrencies.

Table 4.1 The enablers of blockchain

Enablers of blockchain			
Social influence	*Facilitating conditions*	*Effort expectancy*	*Performance expectancy*
• Digital asset adoption • Management and strategy	• Macro drivers • Organisational readiness	• Regulations • Volatility and trust • Education and training	• Direct • Transparent • Immutable • Computational logic

4.3 EXAMINING THE METHODOLOGY AND EXISTING FRAMEWORKS OF ADOPTION MODELS FOR DIGITAL ASSETS

Adoption theory has developed by embracing a range of perspectives and incorporating diverse models. One such model is the self-determination theory (Ryan & Deci, 2000), which underscores individuals' aspirations to take charge and exert influence over their lives when making choices. Prospect Theory (Kahneman & Tversky, 1977) explores people's tendency to embrace risk, while the Theory of Planned Behaviour (Ajzen & Fishbein, 1969) identifies attitudes that mould behavioural intentions. Also, Bandura's (1977) social learning theory and Roger's (1983) diffusion of innovation curve offer broader concepts of innovation adoption.

Nevertheless, several models and frameworks have been developed to explain user adoption of new technologies (Taherdoost, 2018; Venkatesh, 2012). However, the Unified Theory of Acceptance and Use of Technology (UTAUT) model has been widely applied in various technology scenarios. UTAUT has been applied to both industrial and consumer settings. Inferences derived from the UTAUT literature emphasised the need for future research to address novel focal phenomena. The objective is to concentrate on contextual and individual variations, giving explicit consideration to emerging exogenous, endogenous, moderating, outcome, mediating, external, and internal mechanisms. Therefore, in this chapter, we employ the UTAUT framework to explore issues associated with blockchain and digital assets. Therefore, we employ the UTAUT model as a foundational framework in our comprehensive literature review, given its widespread application across various subjects.

Some examples of subjects explored in this study are shown in Table 4.2.

Table 4.2 Examples of subjects explored by UTAUT in recent times

Technologies	References
Mobile applications, internet and payments	Baabdullah et al., (2015), Baabdullah et al., (2019), Slade et al., 2015a,b), Bere, (2014), Choudrie et al., (2014)
Social networks	Herrero et al., (2017)
Banking	Alalwan et al., (2015), Alalwan et al., (2016), Alalwan et al., (2017)
Chatbots	Alsharhan et al., (2024)
Software development	Balaid et al., (2014)
Smartwatches	Chuah et al., (2016)
Edtech	Montes de Oca and Nistor, (2014),
E-government	Dwivedi and Williams, (2008)
Airline tickets	Escobar-Rodríguez and Carvajal-Trujillo, (2013)
Healthcare wearables	Gao et al., (2015)

Digital assets and organisational capabilities 75

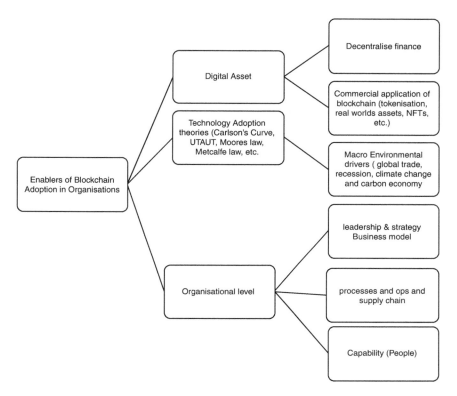

Figure 4.3 Enablers of blockchain adoption in organisations.

The UTAUT model has four key dimensions. Namely, expectation of performance, expectation of effort, social influences, and facilitating conditions (Venkatesh et al., 2012). In recent research, significant mediators and moderators were added to the model. These include personalisation benefits (Krishnaraju et al., 2016), perceived enjoyment (Koenig-Lewis et al., 2015), trust (Slade et al., 2015a,b), behavioural intentions, and the relationship with continuous usage (Wong et al., 2020). Emphasis has also been placed on readiness (Chhonker et al., 2017) and developing digital configuration capabilities (Fakhreddin, 2023). Our framework is provided in Figure 4.3.

To understand the facilitation of blockchain adoption, it is essential to inquire in greater depth the core aspects for acceptance and use, these include performance expectancy, effort expectancy, social influence, and facilitating conditions as seen in Figure 4.4. Figure 4.4 shows the key elements for the acceptance and use of blockchain technology.

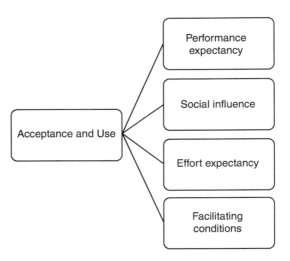

Figure 4.4 The UTAUT foundation framework.

The UTAUT key elements and questions for the acceptance and use of blockchain technology include:

1. Performance expectancy:
    i. What advantages will organisations gain from implementing blockchain?
    ii. What opportunities exist for innovating new services and processes that enhance efficiency and customer acquisition?
    iii. What are the expectations and potential future applications of blockchain?
    iv. Will blockchain contribute to new revenue streams or cost reduction?
    v. In what ways will it enhance overall business performance?
2. Effort expectancy:
    i. How formidable are the challenges associated with adoption?
    ii. What needs to be developed or established to facilitate seamless integration into business systems?
3. Social influence:
    i. How is the current understanding of blockchain technology?
    ii. What are the perceptions of senior leaders and colleagues regarding effective use cases and opportunities for blockchain?
4. Facilitating conditions:
    i. Is the readiness of organisations to adopt blockchain adequate?
    ii. Is there sufficient expertise and training available? Is the wider business community and infrastructure prepared for blockchain implementation?

4.4 FINDINGS AND DISCUSSION

4.4.1 What factors contribute to the facilitation of blockchain enablers

The subsequent section of this chapter covers the essential areas drawn from the literature review that will promote the widespread adoption of blockchain technologies. Initially, we have underscored the importance of digital assets in popularising and catalysing innovation within blockchain. Cryptographic coins (crypto) have, in various respects, evolved into a social influence, fostering conditions conducive to broader adoption of blockchain. Performance Expectancy of Blockchain involves the fundamental principles governing the operation of blockchain, emphasising distinctions between blockchain and traditional technologies (Lansiti & Lakhani, 2017). The distributed database/ledger grants each participant in the chain access to the entire database, containing a thorough history of transactions. Every node or party can autonomously manage and authenticate transaction records, eliminating the need for intermediaries and ensuring decentralised control over information and data.

The blockchain technology would also resolve the problem of record keeping and transparency in the supply chain for customers. Customers are increasingly demanding details for transactions and the source of the manufactured products, including raw materials, suppliers, etc. Track and trace of all kinds of transactions would create efficient supply chains as well as more transparency and security. Therefore, there are lots of opportunities to adopt blockchain technology in the supply chain (Maroun & Daniel 2019, Daniel 2021).

4.4.2 The Macro drivers of blockchain and crypto adoption

The speed of technological advancements is poised to eventually mitigate potential energy consumption concerns and enhance mining efficiency and transaction speed through the adoption of innovative technologies. These may encompass Quantum Computing, Graphene processors, Optical computing, Artificial Intelligence (AI), faster broadband (5G), and edge computing, among others. The impacts of these advancements are manifested in four discernible macro drivers of digital technology and are anticipated to significantly influence the success of blockchain adoption. Table 4.2 is a review of adoption theories within the advancement of technology which consider macro drivers that collectively contribute to the ongoing growth and adoption of blockchain technology and crypto.

Table 4.3 Technology adoption theories

Theory	Description
Moore's Law (1998)	Moore's Law predicts microprocessors double in power every two years, advancing faster and cheaper technology to manage increased data processing in distributed ledgers.
Carlson's Curve (2003)	Calson's Curve shows decreasing DNA decoding time, reflecting data analysis capacity and innovation. Swift processing is crucial for earning rewards in Bitcoin's proof-of-work protocol.
Wright's law (1936) and experience curve effects	Wright's Law ties increased production to lower costs and better performance, while Bitcoin miners improve efficiency with ASIC computers and renewable energy sources.
Henry Adam's Curve (Hall, 2021)	Increased energy drives human growth. Bitcoin and DeFi tech decentralise banking, reducing intermediary reliance. Bitcoin mining incentivises cheap energy use, from renewables or fossil fuels, to minimise the environmental impact.
Metcalfe's Law (Peterson, 2018)	Network value grows exponentially with user base, as seen with fax machines and social networks like Facebook and Google. Blockchain and cryptocurrency similarly promise new use cases and utility over time, driven by investments spurring rapid adoption and innovation.

4.4.3 The strategy and managerial aspects of cryptocurrency adoption from a social influence perspective

A cryptocurrency implementation and adoption strategy involves numerous managerial aspects to ensure successful integration within a business or organisation. The key managerial factors for the application of a successful cryptocurrency adoption strategy feature in Table 4.3.

In essence, businesses rely on trust for successful cryptocurrency integration, which is considered the main variable for adoption in the modern world (Albayati et al., 2020). Education and clear policies drive cryptocurrency adoption, fostering trust (Pastory & Mahwera, 2022). Cryptocurrencies' resistance to inflation and fast, low-cost transactions also support adoption (Wu et al., 2022).

Table 4.4 The strategy and managerial aspects for successful cryptocurrency adoption

No.	Managerial requirements
1	Clear goals are crucial for cryptocurrency adoption, including exploring new business models for faster transactions, cost reduction, or customer base expansion. Well-educated managers typically exhibit stronger expertise and innovation (Kearney et al., 2002).
2	Assess risks, compliance, and regulations for cryptocurrency integration. Mitigate potential risks, anticipate challenges, and stay informed about regulatory environments and standards to address adoption hurdles (Abramova & Bohme, 2016)
3	Evaluate technology infrastructure for cryptocurrency integration, investing in secure frameworks like wallets and payment gateways. Align adoption strategy with operational framework, considering innovation characteristics (Vagnani & Volpe, 2017).
4	Offer educational resources to customers, reliable customer support, and seek collaborations with established cryptocurrency service providers to enhance adoption. Trust is crucial in emerging technologies (Walsh et al., 2021).
5	Develop a financial plan and target audience for cryptocurrency adoption. Manage budgets, accounting, and tax implications with expert advice. Emphasise transparency and continuous communication of benefits and risks in the marketing strategy. Research indicates larger enterprises adopt blockchain technology quicker than smaller businesses (Clohessy & Acton, 2019).
6	Continuously monitor system effectiveness and adapt to evolving cryptocurrency landscapes. Ensure adoption strategy aligns with environmental principles and addresses mining practices and illicit activities (Arpaci, 2023).

4.4.4 Organisational readiness and cryptocurrency from a micro level facilitating conditions perspective

Organisational readiness for blockchain and cryptocurrency refers to the preparedness of a company, institution, or entity to effectively adopt and utilise cryptocurrency-related technologies in its operations. Numerous factors influence an organisation's readiness to engage with blockchain technologies (Chavalala et al., 2022). Companies must dedicate time to understanding digital transformation and emerging technologies (Machado et al., 2021). Historical data suggests that 90% of new ideas fail to materialise due to insufficient organisational readiness (Lokugea et al., 2019).

Engaging skilled consultants who are experts in cryptocurrency adoption can prove immensely advantageous in navigating the intricacies of blockchain (Xia et al., 2022). Organisational preparedness varies based on industry, company size, and specific goals. Nonetheless, research indicates that larger corporations exhibit a higher inclination toward adopting blockchain technology compared to small- or medium-sized enterprises

(Clohessy & Acton, 2019). Prior to organisations commencing cryptocurrency initiatives, conducting a thorough evaluation of readiness which considers essential factors is necessary (Machado et al., 2021), example factors are noted below:

- Strategic management and leadership support: Ensuring that adoption aligns with the overarching business strategy, goals, and objectives. The leadership team's understanding of both the potential benefits and risks remains pivotal. Endorsing a strategic vision for the integration of crypto-related initiatives is a vital element of success (Xia et al., 2022).
- Technical infrastructure and regulatory compliance: Evaluating and enhancing technical infrastructure to securely accommodate cryptocurrency transactions is paramount. This involves implementing resilient security measures to safeguard against cyber threats. Companies must also adjust to legal obligations and stay abreast of evolving regulations to ensure compliance (Bolstler, 2019; Saiedi et al., 2021).
- Employee education, training, and risk management: Offering extensive education and training to employees. This ensures that the workforce is proficient in managing crypto-related responsibilities effectively. Formulating risk management strategies tailored to handle inherent cryptocurrency volatility, security vulnerabilities, and operational risks is crucial for minimising potential downsides (Saiedi et al., 2021; Baur et al., 2018).
- Customer/client education, partnerships, and alliances: Educating customers or clients about cryptocurrency-related services, when relevant, can encourage adoption and strengthen confidence in the organisation's competencies. Partnering with established entities in the cryptocurrency sphere or creating alliances can yield valuable insights, resource access, and support in navigating the complexities of the crypto industry (Machado et al., 2021). Cryptocurrency, including Bitcoin, embodies an innovative and technologically advanced alternative for an uncertain, globalised future. If regulated efficiently, this could assist future generations in handling the challenges related to diverse forms of financial transactions (Afzalur & Dawood, 2019).
- Sustainable pilot projects and testing: Engaging in pilot projects or trials to assess the feasibility, scalability, and potential challenges of integrating cryptocurrencies within the organisation can provide valuable insights before undertaking widespread implementation. Staying adaptable to changes and embracing innovation is crucial in navigating the rapidly evolving landscape of cryptocurrencies. Furthermore, the financial sustainability of cryptocurrencies stands to revolutionise conventional financial systems by enabling faster, cost-effective, and inclusive value transfers in a more environmentally friendly global economy (Arpaci, 2023).

4.5 CONCLUSION

Cryptocurrencies have become increasingly popular in recent years, serving as catalysts for the advancement of blockchain development across various commercial contexts. This chapter brought together from a comprehensive literature review the key aspects of adoption that would enable wider use of blockchain technology. The enablers of blockchain adoption in organisations from digital assets and organisational capabilities are proposed. This chapter used the UTAUT to categorise and order the key enablers into four broad areas of interest. These included: the impact of social influences and facilitating conditions, along with individuals' perceptions of effort expectancy and performance expectancy. Blockchain technology is a pervasive and disruptive technology, which will have the potential to improve the way we organise and manage commercial ventures in the future. Cryptocurrencies have brought blockchain into the public conscience and created a commercial platform for ventures to initiate blockchains. Nonetheless, realising the true potential of blockchain is accompanied by challenges that must be addressed. Establishing a regulatory framework for cryptocurrencies is essential to attract increased seed and institutional funding. Training and education efforts, focusing on both the strategic advantages and technological hurdles, are still in early stages of development. Both academics and industry practitioners have significant roles to play in supporting the sector and facilitating the broader adoption of blockchain technology.

REFERENCES

Abdeldayem, M., & Aldulaimi, S. (2023). Investment decisions determinants in the GCC cryptocurrency market: A behavioural finance perspective, *International Journal of Organizational Analysis*, Vol. ahead-of-print No. ahead-of-print. https://doi.org/10.1108/IJOA-02-2023-3623.

Abramova, S., & Bohme, R. (2016). Perceived benefit and risk as multidimensional determinants of bitcoin use: A quantitative exploratory study. In *Thirty-Seventh International Conference on Information Systems*. Dublin.

Afzalur, R., & Dawood, A.K. (2019). Bitcoin and future of cryptocurrency. *Ushus-Journal of Business Management* 2019, 18(1), 61–66. ISSN 0975-3311. https://doi:10.12725/ujbm.46.5.

Alalwan, A. A., Dwivedi, Y. K., & Rana, N. P. (2017). Factors influencing adoption of mobile banking by Jordanian bank customers: Extending UTAUT2 with trust. *International Journal of Information Management*, 37, 99–110.

Alalwan, A. A., Dwivedi, Y. K., Rana, N. P., Lla, B., & Williams, M. D. (2015). Consumer adoption of Internet banking in Jordan: Examining the role of hedonic motivation, habit, self-efficacy and trust. *Journal of Financial Services Marketing*, 20, 145–157.

Alalwan, A. A., Dwivedi, Y. K., & Williams, M. D. (2016). Customers' intention and adoption of telebanking in Jordan. *Information Systems Management*, 33, 154–178.

Alabi, K. (2017). Digital blockchain networks appear to be following Metcalfe's Law. *Electronic Commerce Research and Applications*, 24, 23–29.

Albayati, H., Kim, S. K., & Rho, J. J. (2020). Accepting financial transactions using blockchain technology and cryptocurrency: A customer perspective approach. *Technology in Society*, 62(101320), 1–20.

Albrecht, C., Hawkins, S., & Duffin, K.M. (2020). Legitimizing bitcoin as a currency and store of value: using discrete monetary units to consolidate value and drive market growth. *LEDGER*, 5 (1), 1–10. https://doi.org/10.5915/LEDGER.2020.167.

Al-Mashari, M., Zairi, M., & Okazawa, K. (2006). Enterprise Resource Planning (ERP) implementation: a useful road map. *International Journal of Management and Enterprise Development*, 3(1–2), 169–180.

AlShamsi, M., Al-Emran, M., & Shaalan, K. (2022). A systematic review on blockchain adoption. *Applied Sciences*, 12(9), 4245. https://doi.org/10.3390/app12094245

Alsharhan, A., Al-Emran, M., & Shaalan, K. (2024). Chatbot adoption: A multiperspective systematic review and future research agenda. *IEEE Transactions on Engineering Management*, 71, 10232–10244. doi: 10.1109/TEM.2023.3298360.

Ajzen, I., & Fishbien, M. (1969). The prediction of behavioural intentions in a choice situation. *Journal of Experimental Social Psychology*, 5(4), 400–416. https://doi.org/10.1016/0022-1031(69)90033-X.

Arpaci, I. (2023). Predictors of financial sustainability for cryptocurrencies: An empirical study using a hybrid SEM-ANN approach. *Technological Forecasting & Social Change*, 196, 122858.

Bandura, A. (1977). Self-efficacy: Toward a unifying theory of behavioral change. *Psychological Review*, 84(2), 191–215. https://doi.org/10.1037/0033-295X.84.2.191.

Baabdullah, A., Dwivedi, Y., Williams, M., & Kumar, P. (2015). *Understanding the adoption of mobile internet in the Saudi Arabian context: Results from a descriptive analysis*. Springer International Publishing.

Baabdullah, A. M., Alalwan, A. A., Rana, N. P., Kizgin, H., & Patil, P. (2019). Consumer use of mobile banking (M-Banking) in Saudi Arabia: Towards an integrated model. *International Journal of Information Management*, 44, 38–52.

Baur, D. G., Dimpfl, T., & Kuck, K. (2018). Bitcoin, gold and the US dollar – a replication and extension. *Finance Research Letters*, 25, 103–110. https://doi.org/10.1016/J.FRL.2017.10.012.

Balaid, A., .Rozan, M. Z., & Adullah, S. N. (2014). Conceptual model for examining knowledge maps adoption in software development organizations. *Asian Social Science*, 10, 118.

Bere, A. (2014) Exploring determinants for mobile learning user acceptance and use: An application of UTAUT. 2014 *11th International Conference on Information Technology: New Generations*, 2014. IEEE, 84–90. Las Vegas, NV, United States.

Bolstler, M. (2019) The influence of cryptocurrencies on enterprise risk management – An empirical evidence by the example of bitcoin. *Junior Management Science*, 4(2), 195–227.

Carlson, R. (2003). The pace and proliferation of biological technologies. Biosecurity and Bioterrorism: Biodefense Strategy, Practice, *and Science, 1*(3), 203–214.

Chavalala, M.M., Bag, S., Pretorius, J.H.C., & Rahman, M.S. (2022), A multi-method study on the barriers of the blockchain technology application in the cold supply chains, *Journal of Enterprise Information Management*. https://doi.org/10.1108/JEIM-06-2022-0209.

Chuen, D. L. K., Guo, L., & Wang, Y. (2018). Cryptocurrency: A new investment opportunity? *The Journal of Alternative Investments, 20*(3), 16–40. https://doi.org/10.3905/jai.2018.20.3.016.

Choudrie, J., Pheeraphuttharangkoon, S., Zamani, E., & Giaglis, G. (2014). Investigating the adoption and use of smartphones in the UK: a silver-surfers perspective. In *22nd European Conference on Information Systems, ECIS 2014*. Association for Information Systems.

Chhonker, M. S., Verma, D., & Kar, A. K. (2017). Review of technology adoption frameworks in mobile commerce. *Procedia Computer Science, 122*, 888–895.

Chuah, S. H.-W., Rauschnabel, P. A., Krey, N., Nguyen, B., Ramayah, T., & Lade, S. 2016. Wearable technologies: The role of usefulness and visibility in smartwatch adoption. *Computers in Human Behavior, 65*, 276–284.

Clohessy, T., & Acton, T. (2019). Investigating the influence of organizational factors on blockchain adoption: An innovation theory perspective. *Industrial Management & Data Systems, 119*(7), 1457–1491.

Daniel, J. (2021). Blockchain Technology for Businesses: Trends and Research Themes. *European Decision Sciences Institute Conference, Decision Sciences Institute.*

Dourado, Eli, & Jerry, B. (2014) In the New Palgrave Dictionary of Economics, retrieved from https://www.researchgate.net/publication/298792075_Cryptocurrency on 6 October 2018.

Duda, M. (2015). The outline of the history of money development and of the monetary system. Selected aspects of the issue. *Wspolczesna Gospodarka, 6*(2), 9–20.

Dwivedi, Y. K., & Williams, M. D. (2008). Demographic influence on UK citizens'e-government adoption. *Electronic Government, an International Journal, 5*, 261–274

EPRS (2023). Non-EU countries' regulations on crypto-assets and their potential implications for the EU. *European Parliamentary Research* Service (EPRS). EPRS_BRI(2023)753930_EN.pdf.

Escobar-Rodriguez, T., & Carvajal-Trujillo, E. (2013). Online drivers of consumer purchase of website airline tickets. *Journal of Air Transport Management, 32*, 58–64.

Fakhreddin, F. (2023). 5 How digital technology adoption results in improved innovation and firm performance outcomes. *Business Digitalization: Corporate Identity and Reputation, 63*(1), 63–75.

Gao, Y., LI, H., & Luo, Y. (2015). An empirical study of wearable technology acceptance in healthcare. *Industrial Management & Data Systems, 115*, 1704–1723.

Gupta M, & Sharma A. (2021). Fear of missing out: A brief overview of origin, theoretical underpinnings and relationship with mental health. *World Journal of Clinical Cases, 9*(19), 4881–4889. https://doi.org/10.12998/wjcc.v9.i19.4881. PMID: 34307542; PMCID: PMC8283615.

Gunasekaran, A., Patel, C., & McGaughey, E. (2004). A Framework for supply chain performance measurement. *International Journal of Production Economics*, 87(3), 333–347. https://doi.org/10.1016/j.ijpe.2003.08.003

Hall, J. S. (2021). *Where Is My Flying Car*. Stripe Press.

Haynes, A., & Yeoh, A. (2018). *Cryptocurrencies and Cryptoassets*. Regulatory and Legal Issues. Informa Law. Routledge.

Herman, D. (2000). Introducing short-term brands: A new branding tool for a new consumer reality. Journal of Brand *Management*, 7(5), 330–340.

Herrero, Á., San Martin, H., & Garcia-De Los Salmones, M. D. M. (2017). Explaining the adoption of social networks sites for sharing user-generated content: A revision of the UTAUT2. *Computers in Human Behavior*, 71, 209–217.

Iansiti, M., & K. R. Lakhani (2017). The truth about blockchain. *Harvard Business Review*, 95(1), 118–127.

Johnson, S. (2018). *Beyond the Bitcoin Bubble*. The New York Times.

Kearney, R. C., Feldman, B. M., & Scavo, C. P. F. (2002) Reinventing government: City Manager Attitudes and Actions. *Public Administration Review*. 60(6). https://doi.org/10.1111/0033-3352.00116

Laurence, T. (2023). *Blockchain for Dummies*. Third Edition. John Wiley and Sons.

Lacity, M. C. (2018). Addressing key challenges to making enterprise blockchain applications a reality. *MIS Quarterly Executive* 17(3): 201–222.

Learning Curves & Wright's Law. (n.d.) Understanding what drives exponential... | by Tom Connor | 10x Curiosity | Medium. https://medium.com/10x-curiosity/learning-curves-and-wrights-law-744b85b897a2

Lokugea, S., Sederaa, D., Groverb, V., & Xuc, D. (2019). Organizational readiness for digital innovation: Development and empirical calibration of a construct. *Information & Management*, 56(3), 445–461.

Li, T., Shin, D., & Wang, B. (2021). Cryptocurrency pump-and-dump schemes (February 10, 2021). Available at SSRN: https://ssrn.com/abstract=3267041 or http://dx.doi.org/10.2139/ssrn.3267041.

Krishnaraju, V., Mathew, S.K., & Sugumaran, V. (2016). Web personalization for user acceptance of technology: An empirical investigation of E-government services. *Information Systems Frontiers*, 18, 579–595, https://doi.org/10.1007/s10796-015-9550-9

Machado, C.G., Almstrom, P., Almstrom, M.W., Oberg, A.E., Kurdve, M., & AlMashalah, S. (2021). Digital organisational readiness: Experiences from manufacturing companies. *Journal of Manufacturing Technology Management*, 32(9), 167–182.

Maroun, E. A., & Daniel, J. (2019). Opportunities for use of blockchain technology in supply chains: Australian manufacturer case study. *Proceedings of the International Conference on Industrial Engineering and Operations Management*. Bangkok, Thailand.

Montes De Oca, A.M., & Nistor, N. (2014). Non-significant intention–behavior effects in educational technology acceptance: A case of competing cognitive scripts? *Computers in Human Behavior*, 34, 333–338.

Moore, G. E. (1998). Cramming more components onto integrated circuits. *Proceedings of the IEEE*, 86(1), 82–85.

Nakamoto, S. (2008). Bitcoin: A peer-to-peer electronic cash system. Retrieved November from www.cryptovest.co.uk

Notheisen, B., Hawlitschek, F., & Weinhardt, C. (2017). Breaking down the blockchain hype–towards a blockchain market engineering approach. *In Proceedings of the 25th European Conference on Information Systems (ECIS)*, Guimarães, Portugal, June 5–10, 2017 (pp. 1062–1080). ISBN 978-989-20-7655-3 Research Papers. http://aisel.aisnet.org/ecis2017_rp/69

Pastory, D., & Mahwera, D. A. (2022). Financial institutions readiness towards cryptocurrency adoption: A case of banks in Tanzania. *Journal of Business and Management Review*, 3(10). 740–753. https://doi.org/10.47153/jbmr310.4672022

Peterson, T. (2018). Metcalfe's Law as a Model for Bitcoin's Value. *Alternative Investment Analyst Review Q*, 2.

Popper, N. (2015). Digital gold: Bitcoin and the inside story of the misfits and millionaires trying to reinvent money. Editionunabridged, HarperCollins, 2016, ISBN0062572067, 9780062572066, p. 435.

Przybylski, A. K., Murayama, K., DeHaan, C. R., & Gladwell, V. (2013). Motivational, emotional, and behavioral correlates of fear of missing out. *Computers in Human Behavior*, 29(4), 1841–1848.

Rahman, Z. (2003). Internet-based supply chain management: using the Internet to revolutionize your business. *International Journal of Information Management*, 23(6), 493–505.

Rogers, E.M. (1983). *Diffusion of Innovations*. Third Edition. The Free Press A Division of Macmillan Publishing Co., Inc

Ruangkanjanases, A., Hariguna, T., Adiandari, A. M., & Alfawaz, K. M. (2022). Assessing blockchain adoption in supply chain management, antecedent of technology readiness, knowledge sharing and trading need. *Emerging Science Journal*, 6, 921–937.

Ryan, R. M., & Deci, E. L. (2006). Self-regulation and the problem of human autonomy: Does psychology need choice, self-determination, and will? *Journal of Personality*, 74(6), 1557–1586. https://doi.org/10.1111/j.1467-6494.2006.00420.x

Saiedi, E., Broström, A. & Ruiz, F. (2021). Global drivers of cryptocurrency infrastructure adoption. *Small Business Economics* 57, 353–406. https://doi.org/10.1007/s11187-019-00309-8.

Shiller, R. J. (2014). Speculative *asset prices. American Economic Review*, 104(6), 1486–1517. https://doi.org/10.1257/aer.104.6.1486.

Shiller, R. J. (2015). Irrational exuberance. In *Irrational Exuberance*. Princeton University Press. ISBN1400865530, 9781400865536, p. 392.

Shiller, R (2023) *Bitcoin's value is tied to its Narrative.* Shortform. https://www.shortform.com/blog/robert-shiller-bitcoin/

Slade, E., Williams, M., Dwivdei, Y. & Piercy, N. (2015a). Exploring consumer adoption of proximity mobile payments. *Journal of Strategic Marketing*, 23(3), 209–223. https://doi.org/10.1080/0965254X.2014.914075

Slade, E., Dwivedi, Y.K, Piercy, N.C., & Williams, M.D. (2015b). Modelling consumers' adoption intentions of remote mobile payments in the United Kingdom: Extending UTAUT with Innovativeness, risk, and trust. *Psychology of Marketing*, 32(8), 860–873. https://doi.org/10.1002/mar.20823.

Statista (2023). www.statista.com https://www.statista.com/?kw=statista&crmtag=adwords&gclid=EAIaIQobChMIoZGqv5HtgwMVWZdQBh3DuQEhEAAYASAAEgJT8_D_BwE

Sukumaran, S., Bee, T. S., & Wasiuzzaman, S. (2022). Investment in cryptocurrency: A study of it adoption among Malaysian investors. *Journal of Decision Systems*, 32(4), 732–760. https://doi.org/10.1080/12460125.2022.2123086.

Taherdoost, H. (2018). A review of technology acceptance and adoption models and theories. *Procedia Manufacturing*, 22, 960–967, https://doi.org/10.1016/j.promfg.2018.03.137.

Tversky, A., & Kahneman, D. (1977). Causal thinking in judgment under uncertainty. In R. Butts & J. Hintekka (Eds.), *Basic Problems in Methodology and Linguistics* (pp. 167–190). Dordrecht: D. Reichel Publishing Company.

Venkatesh, V., Thong, J. Y., & Xu, X. (2012). Consumer acceptance and use of information technology: Extending the unified theory of acceptance and use of technology. *MIS Quarterly*, 157–178.

Venkatesh, V., Thong, J. Y., & Xu, X. (2016). Unified theory of acceptance and use of technology: A synthesis and the road ahead. *Journal of the association for Information Systems*, 17(5), 328–376.

Vagnani, G., & Volpe, L. (2017). Innovation attributes and managers' decisions about the adoption of innovations in organizations: A meta-analytical review. *International Journal of Innovation Studies*, 1(2), 107–133.

Vicente, D. M., Duarte, P. D., & Granadeiro, C. (2023). Fintech regulation and the licensing principle, *centro de Investigaçao de direito privado, European Banking Institute, Frankfurt, Bocconi Legal Studies Research Paper* No. 4346795. https://ebi-europa.eu/wp-content/uploads/2023/02/eBook-22Fintech-Regulation-Licensing-Principle-2-2023.pdf

Walsh, C., O'Reilly, P., Gleasure, R., McAvoy, J., & O'Leary, K. (2021). Understanding manager resistance to blockchain systems. *European Management Journal*, 39(3), 353–365.

Wong, L. W., Tan, G. W. H., Lee, V. H., Ooi, K. B., & Sohal, A. (2020). Unearthing the determinants of blockchain adoption in supply chain management. *International Journal of Production Research*, 58(7), 2100–2123.

Wright, T. P. (1936). Factors affecting the cost of airplanes. Journal of the Aeronautical Sciences, 3(4), 122–128.

Wu, R., Ishfaq, K., Hussain, S., Asmi, F., Siddiquei, A. N., & Anwar, M. A. (2022). Investigating e-retailers' intentions to adopt cryptocurrency considering the mediation of technostress and technology involvement. *Sustainability*, 14(641), 1–21.

Xia, C., Miraz, M. M., Gazi, A.I., Rahaman, A., Habib, M., & Hossain, A.I. (2022). Factors affecting cryptocurrency adoption in digital business transactions: The mediating role of customer satisfaction. *Technology in Society*, 70, 102059.

Yilmaz, N.K., & Hazar, H.B. 2018). Predicting future cryptocurrency investment trends by conjoint analysis. *Press Academia*, 5(1), 321–330. https://doi.org/10.17261/Pressacademia.2018.999.

Zhao, H., & Zhang, L. (2021). Financial literacy or investment experience: which is more influential in cryptocurrency investment? *International Journal of Bank Marketing*, 39(7), 1208–1226. https://doi.org/101108/IJBM-11-2020-0552.

Part 3

Blockchain adoption and enabling technologies

Chapter 5

Blockchain technology and artificial intelligence for enhanced vaccine supply chain management

Utkarsh Mittal and Amit Kumar Yadav

5.1 INTRODUCTION

Influenza vaccination is a key public health priority for the Centers for Disease Control and Prevention (CDC), which strives to expand access and reduce disparities. However, mismatches between flu vaccine supply and demand persist, undermining immunization goals (Samii *et al.* 2012). During the 2020–2021 season, despite CDC stockpiling over 11 million supplementary doses, overall vaccination rates increased only marginally[1]. More concerningly, coverage dropped among high-risk groups like children and minorities. Racial and ethnic disparities in vaccine uptake also endured, though these communities face elevated hospitalization risks from influenza (Na *et al.* 2023).

Such shortfalls highlight systemic deficiencies in aligning vaccine production and allocation with public health needs. Demand variability across seasons and populations exacerbates uncertainties in forecasting and distribution planning. Consequently, inequities arise in vaccine availability and accessibility. There is an urgent need to enhance analytical capabilities for optimized, equitable vaccine supply chain management with better transparency and traceability across network.

Vaccination is a vital public health intervention that plays a crucial role in preventing the spread of infectious diseases and safeguarding global health (Adida *et al.* 2013). However, deficiencies in vaccine supply chains, such as unreliable demand forecasts, quality control issues, and coordination difficulties, hinder access to vaccines and impede the achievement of immunization goals (Chandra & Kumar 2018, Duijzer *et al.* 2018). The COVID-19 pandemic has further exacerbated vaccine supply challenges, resulting in shortages and delivery failures (Bloom *et al.* 2021). Consequently, there is an urgent need for a sophisticated vaccine supply chain management system.

Previous research has investigated the application of blockchain for enhancing supply chain transparency (Abeyratne 2016), the utilization of Internet of Things (IoT) for real-time monitoring (Tajima 2007), and the implementation of machine learning (ML) for demand forecasting

(Carbonneau *et al.* 2008). However, studies examining integrated systems that combine these technologies are scarce. Liu et al. (2021) assessed the influence of blockchain on vaccine supply chain pricing and coordination but did not incorporate IoT or ML. Similarly, Hasan et al. (2019) proposed the use of blockchain in container supply chains without focusing on vaccines. The potential benefits of employing emerging technologies in vaccine supply chains have not been adequately explored.

This study develops an intelligent vaccine supply management system (IVSCMS) integrating blockchain, IoT sensors, and ML algorithms to enhance transparency, quality, forecasting, and coordination. The blockchain provides tamper-proof vaccine tracking (Adarsh *et al.* 2021), IoT enables real-time monitoring (Singh *et al.* 2020), and ML analyzes supply chain data for demand projections. Recurrent neural networks such as LSTMs (Long Short-Term Memory) are for vaccine demand forecasting across multiple seasons. These models can uncover complex temporal relationships in seasonal epidemiological datasets. The chapter offers three key contributions. First, it establishes a vaccine supply chain management system synthesizing blockchain, IoT, and ML. Second, it demonstrates this system's effectiveness through simulated testing of the integrated technologies. Third, it provides a digital transformation model for augmenting supply chain performance.

The rest of this chapter is organized as follows: Section 5.2 reviews the impact of COVID-19 and the implementation of vaccine supply chains and emerging technologies. Section 5.3 discusses methodology blockchain, IoT, and ML innovations. Section 5.4 presents ML models. Section 5.5 analyzes simulated vaccine data using ML models. Section 6 discusses the result and its implications, and finally, Section 5.7 concludes the study with future research directions.

5.2 LITERATURE REVIEW

The literature on vaccine distribution and planning has been growing in recent years, with a focus on exploring the potential of blockchain, IoT, and AI technologies in transforming vaccine supply chain management. These technologies have been identified as having the potential to address long-standing challenges such as lapses in the cold chain, unpredictable demand, lack of coordination, and limited transparency (Chandra & Kumar 2018, Duijzer *et al.* 2018).

Several studies have assessed the potential of blockchain technology in improving vaccine supply chains (Chauhan *et al.* 2021, Musamih *et al.* 2021, Yadav *et al.* 2023).

Liu et al. (2021) found that blockchain enhances transparency, fosters trust through immutability, and increases profitability in vaccine pricing and coordination models. Musamih et al. (2021) proposed an Ethereum blockchain architecture for managing COVID-19 vaccine distribution data and logistics. Yong et al. (2020) designed a blockchain-based system called Intelligent Vaccine Management and Supervision System (IVMSS) to ensure vaccine authenticity and safety. Cui et al. (2021) developed a blockchain platform for tracking COVID-19 vaccines from production to injection to improve accountability. However, these studies have focused primarily on blockchain and have not explored its potential integration with IoT and AI.

Researchers have also investigated the potential of IoT for real-time monitoring and quality assurance in vaccine supply chains. Monteleone et al. (2017) implemented IoT sensors in conjunction with RFID tags to monitor cold chain conditions during vaccine transport and storage. Lin et al. (2020) developed an IoT-enabled smart container model that featured temperature controllers and warning systems to prevent cold chain breaches during distribution. However, the full potential of IoT in combination with blockchain and AI has yet to be fully explored.

The application of ML techniques for forecasting vaccine demand has shown great potential in utilizing multivariate historical data. Sarkar et al. (2020) employed long short-term memory neural networks to model the trajectory of COVID-19 spread, with the aim of informing just-in-time vaccine allocation strategies. Despite these advancements, research specifically focused on ML for forecasting vaccine demand remains scarce.

Several studies have suggested the integration of blockchain with IoT and AI, but few have focused specifically on vaccine supply chains. Tsolakis et al. (2021) explored the use of blockchain and IoT to enhance the sustainability of food supply chains. Wang et al. (2021) developed a blockchain IoT architecture for tracking traditional Chinese medicine. Tanwar et al. (2020) investigated the potential of combining blockchain, IoT, and ML to improve supply chain performance. However, the literature reveals a scarcity of research that specifically examines the convergence of these technologies in the context of vaccine supply chains.

During the COVID-19 pandemic, Kraus et al. (2021) explored the potential of digital transformation in healthcare supply chains. Dwivedi et al. (2021) conducted a comprehensive review of the applications of AI techniques, such as ML, in various domains, including supply chain management. However, the research examining the use of transformational AI in vaccine supply chains is currently quite limited. A summary of blockchain applications along with AI and IoT for Vaccine Supply Chains is given in Table 5.1.

Mittal et al. (2008) proposed a methodology rooted in physics-based principles, which leverages nonlinear differential equations and a customized objective function to project a firm's throughput. This approach emphasizes the viability of physics-based strategies in forecasting. Nevertheless, the model's reliance on extensive data preprocessing and its potential limitations in extrapolating to diverse supply chain environments should be acknowledged. The research conducted by Mittal et al. (2008) delved into the complexities and susceptibilities inherent in supply chains, which are

92 Blockchain Technology

Table 5.1 Summary of blockchain applications for vaccine supply chains

Author	Technology	Key takeaways	Limitations	Type of AI technology
Liu et al. (2021)	Blockchain	Enhances transparency, trust, and profitability in vaccine pricing and coordination	Focused only on blockchain, no integration with IoT and AI	
Musamih et al. (2021)	Blockchain (Ethereum)	Proposed architecture for managing COVID-19 vaccine data and logistics	Did not explore integration potential	
Yong et al. (2020)	Blockchain (IVMSS system)	Ensured vaccine authenticity and safety	Narrow focus on blockchain	
Cui et al. (2021)	Blockchain	Vaccine tracking from production to injection for accountability	Limited to blockchain	
Monteleone et al. (2017)	IoT + RFID	Cold chain tracking during vaccine transport and storage	Did not examine convergence with blockchain and AI	
Lin et al. (2020)	IoT (smart containers)	Temperature control and monitoring to prevent cold chain breaches	IoT's full potential not explored	
Sarkar et al. (2020)	Machine learning (LSTM)	COVID-19 spread modeling for vaccine allocation	Focused only on ML for demand forecasting	LSTM neural networks
Tsolakis et al. (2021)	Blockchain + IoT	Food supply chain sustainability	Not focused on vaccines	
Wang et al. (2021)	Blockchain + IoT	Tracking traditional Chinese medicine	Did not examine vaccines	
Tanwar et al. (2020)	Blockchain + IoT + ML	Supply chain performance enhancement	No vaccine supply chain focus	
Kraus et al. (2021)	Digital transformation	Potential in healthcare supply chains	Did not specify vaccines	
Dwivedi et al. (2021)	AI (ML)	Applications across domains including SCM	Minimal focus on vaccine supply chains	Review of ML techniques
Mittal et al. (2008)	Physics-based models	Forecasting throughput with differential equations	Extensive data preprocessing, generalization concerns	Physics-based models
Mittal and Panchal (2023)	AI (ML, DL – CNN)	Predicting supply chain risks and performance	Small dataset size	ML and DL models like CNN

commonly impacted by external disruptions such as pandemics, conflict scenarios, and inflation.

The objective of this study is to develop an AI-driven system capable of accurately assessing these complexities within the domain and effectively mitigating their vulnerabilities. To achieve this, an empirical approach was utilized, incorporating datasets from various studies to develop ML and Deep Learning (DL) models, including linear regression, DL, and CNN networks. These models were designed to predict supply chain risks and enhance the overall stability and performance of an industrial supply chain system. Although the CNN model demonstrated potential, a limitation was the small size of the dataset used.

The integration of blockchain, IoT sensors, and ML analytics has demonstrated value in managing vaccine supply chains independently. However, the research on integrating these technologies into an integrated intelligence system is limited, highlighting a significant gap in the literature. This study aims to address this gap by developing a vaccine supply chain architecture that synthesizes these technologies.

5.3 METHODOLOGY

This research employs a design science methodology to create and assess an artifact: the IVSCMS. The IVSCMS integrates breakthroughs in blockchain, IoT, and ML to tackle practical challenges in vaccine supply chains.

The methodology is outlined in the flowchart given in Figure 5.1.

Initially, we conducted a comprehensive review of existing literature to identify the primary obstacles in the vaccine supply chain and evaluate the potential applications of emerging technologies. Subsequently, we formulated

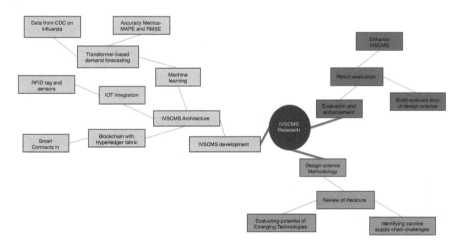

Figure 5.1 Detailed methodology outlines the research.

and developed an innovative IVSCMS architecture that incorporates blockchain, IoT, and ML to effectively address these challenges.

The blockchain platform was developed using Hyperledger Fabric, with smart contracts implemented in chaincode to facilitate vaccine tracking and coordination. Additionally, IoT sensors, such as RFID tags and temperature/humidity sensors, were integrated to gather real-time vaccine monitoring data.

Transformer-based models are utilized for demand forecasting as they can uncover long-term temporal relationships while handling seasonal. These models were trained and assessed on historical data from the CDC regarding influenza vaccinations from 1980 to 2022. The accuracy of the forecasts was determined using standard metrics, including the Mean Absolute Percentage Error (MAPE) and the Root Mean Square Error (RMSE).

The evaluation of the results was undertaken with the aim of enhancing the IVSCMS architecture and rectifying any shortcomings uncovered during the testing phase. This methodological approach aligns with the build-evaluate loop characteristic of design science research, thereby facilitating the creation of an innovative solution to practical challenges.

5.3.1 Implementing blockchain technology for enhanced vaccine management

The proposed blockchain system for vaccine management consists of three integral components: the Good Manufacturing Practice (GMP) chain, the Release chain, and the Inoculation chain. These chains are overseen and maintained by distinct entities – vaccine producers, the lot release agency, and the CDC, respectively. Each vaccine travels through these institutions, and all the associated supply chain data are accurately recorded on the blockchain.

The implementation of smart contracts serves several crucial purposes. Firstly, they ensure comprehensive vaccine traceability, enabling the tracking of each vaccine's journey from production to inoculation. This reduces the risk of misplacement, theft, or misuse of vaccines. Secondly, smart contracts are instrumental in detecting fraud by verifying the authenticity of the vaccines and ensuring that all transactions in the supply chain are legitimate. Lastly, smart contracts allow for easy access to information, permitting both consumers and institutions to retrieve details about inoculation records and the circulation history of the vaccines.

To illustrate, one type of smart contract can verify the production and expiry dates of vaccines to prevent expired vaccines from being administered to patients. Another smart contract variant enables users to inquire about their inoculation records and the circulation history of the vaccines they received, enhancing transparency and boosting confidence in the immunization process.

To further promote accountability within the vaccine supply chain, the system employs a VaccCoin virtual currency. This digital currency, coupled with smart contract payment logic, serves as an incentive for all

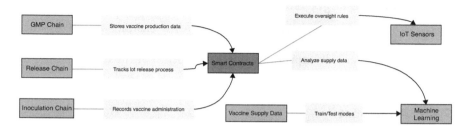

Figure 5.2 System architecture showing key components of the vaccine blockchain and their relationships.

stakeholders to uphold the integrity of the supply chain by aligning economic rewards with responsible behavior. This encourages all participants to prioritize safety, quality, and transparency in their operations. The system architecture of the proposed vaccine blockchain and their relationships is given in Figure 5.2.

5.3.2 Vaccine demand data

The dataset related to vaccine demand forecast offers a comprehensive overview of the total doses of influenza vaccines distributed in the US observed over the last 42 years, from 1980 to 2022. This dataset has been segmented into two parts: the training data, which comprises data from 1980 to 2016, and the testing data, which includes data from 2017 to 2022.

The training data, spanning 37 years, is utilized to develop and refine the predictive models. The extended duration enables the capture of intricate patterns and fluctuations in demand over multiple influenza seasons.

The testing data, covering five years, is employed to evaluate the reliability and accuracy of the predictive models. This data provides an opportunity to compare the model's predictions with actual data, thereby enabling the assessment of the model's performance and facilitating any necessary adjustments.

5.4 MACHINE LEARNING EXPERIMENTS

Multiple experiments were conducted with several ML and DL models to forecast vaccine demand in the supply chain. The models explored include Multiple Regression, AdaBoost, XGBoost, Deep Neural Networks (DNN), Convolutional Neural Networks, and Temporal Fusion Transformers (TFTs). Hyperparameter tuning was performed for the models to improve accuracy on the vaccine dataset.

5.4.1 Multiple regression

First, a multiple linear regression model with a custom objective function as the baseline is employed. However, the model has limitations: it struggled to effectively capture the intricate nature of the data, leading to suboptimal results.

5.4.2 AdaBoost

The efficacy of Adaptive Boosting (AdaBoost) has been investigated. AdaBoost is a commonly used boosting tactic that matches sequences of deficient learners with frequently altered versions of information. The forecasts from these learners were subsequently amalgamated using a pondered total to create a concluding forecast.

The equation $\hat{y} = \text{sign}\left(\sum_{t=1}^{T} W_t f_t(x)\right)$ represents the final prediction, where $\hat{y}$ is the predicted output, T is the number of weak learners, W_t is the weight of the t_{th} weak learner, and $f_t(x)$ is the output of the t_{th} weak learner for the input x.

Initially, the weights ai were set to 1/N, where N is the sample size. The initial phase involves training a weak learner using the original data. In each subsequent iteration, the sample weights were adjusted individually, and the learning algorithm was applied again to the data with updated weights. At any given stage, the weights of the training samples that were inaccurately predicted by the previously induced boosted model increased, whereas the weights for correctly predicted examples decreased. Specifically,

$$a_i = a_i e^{-W_t} \quad \text{if } f_t(x_i) = y_i \qquad (5.1)$$

$$a_i = a_i e^{W_t} \quad \text{if } f_t(x_i) \neq y_i \qquad (5.2)$$

Then the coefficient W_t is updated by $0.5 \ln\left(\dfrac{1 - \text{weighted error}(f_t)}{\text{weighted error}(f_t)}\right)$, where W_{te} is the weighted error.

In the experiment, we used the default setting from scikit-learn, in which the number of weak learners was 50 and the learning rate was 1.

5.4.3 XGBoost

The XGBoost framework applied in the experiment is regarded for its scalability and computational efficiency, which are achieved by optimizing specific loss functions and incorporating regularization techniques. The algorithm has various optimization techniques in place, such as a unique tree-based learning algorithm designed for sparse data, and a weighted quantile sketch method used to manage the instance weights. Moreover, XGBoost utilizes parallel and distributed computing, which accelerates the learning process and enables rapid model exploration. $L(\varphi)$ serves as a loss function that gauges the variation between the prediction and the target for each instance. However, Ω acts as a penalty term that discourages an increase in model complexity.

The hyperparameters being tuned include estimators, which take values between 100 and 500, max depth ranging from 3 to 8, learning rate varying from 0.01 to 0.3, subsample ranging from 0.5 to 1.0, Col sample by tree ranging from 0.5 to 1.0, and gamma with values of 0, 0.25, 0.5, and 1.0. The Randomized SearchCV class was utilized to perform the search, with 50 iterations and 5-fold cross-validation. The evaluation metric used was accuracy. After fitting the randomized search object to the training data, the best parameters were stored in the best parameters.

5.4.4 Deep neural network

The model architecture consists of a feedforward neural network built using the Keras Sequential API. It consists of multiple layers connected sequentially, and the input layer is defined using the dense function with 128 units, a normal distribution-based kernel initializer, and the ReLU activation function. The number of input units has been determined based on the shape of the training data with three hidden layers in the model, each containing 256 units. These layers also use a normal distribution-based kernel initializer and ReLU activation function, and the output layer has a single unit that predicts a continuous value. It uses a normal distribution-based kernel initializer and linear activation function.

5.4.5 Deep convolutional neural network with pool layer

The model design consists of several layers, starting with an embedding layer. This is followed by a convolution layer that uses 128 filters or kernels of size three. A max-pooling layer is applied to each layer of the network. The input data are processed through a sequence of three convolutional and three max-pooling operations. In the final pooling layer, global max pooling is used, which features two compact layers designed with a dropout rate of 0.2 to prevent overfitting. The activation function for these layers is linear, chosen for its ability to produce a continuous output for regression tasks, and is divided into three parts: training (80%), validation (10%), and testing (10%). The experimental setup included a batch size of 128 and an RMSProp optimizer with a learning rate of 0.01. A dropout rate of 20% has been applied where needed. The number of epochs gradually increased from 10 to 100.

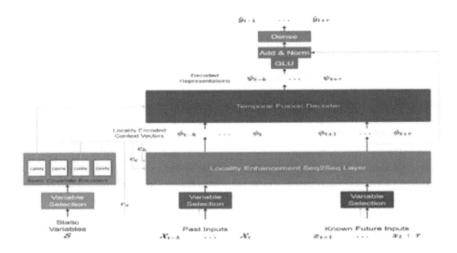

Figure 5.3 Temporal fusion transformer architecture (TFT).

5.4.6 Temporal fusion transformer

A TFT model is incorporated into the blockchain system to analyze the time-series vaccine supply chain data and make accurate demand forecasts. The gating mechanism of TFT customizes the model complexity based on the vaccine dataset. The variable selection identifies the most relevant supply chain factors like production rate, inventory level, etc. for each time step's forecast. The static covariate encoder assimilates contextual information like vaccine type, geography, demographics, etc. to inform the temporal dynamics. The sequence-to-sequence layers capture local trends and patterns while the multi-head attention extracts long-term dependencies in the supply-demand time series. Finally, the TFT model makes a range of demand forecast values for each prediction interval using quantile prediction. This comprehensive architecture equips the TFT to uncover complex temporal relationships in the vaccine supply chain data on the blockchain system for reliable and accurate demand forecasts (Figure 5.3).

5.5 VACCINE DEMAND FORECASTING IS EVALUATED ON THREE DIFFERENT METRICS

5.5.1 Evaluation metrics

The three error measurements of RMSE, MAPE, and mean absolute error (MAE) are used to evaluate the predictive accuracy of the models. RMSE is computed by taking the square root of the average squared differences between predicted and actual values, and it measures the magnitude of the error. MAPE is the mean of the absolute percentage errors, calculated by taking the absolute difference between each prediction and actual value, dividing by the actual value, and averaging these percentages, and it expresses accuracy as a percentage of the error. Finally, MAE is the mean of the absolute differences between predictions and actuals, providing an absolute measure of the typical magnitude of the errors.

These three metrics offer a comprehensive assessment of the deviations between the models' predicted values and the true empirical values. Analyzing RMSE, MAPE, and MAE enables a robust evaluation of the models' predictive capabilities.

$$\text{MAPE} = \frac{\sum_{t=1}^{k}|\widehat{y}_t - y_t|/y_t}{k} \qquad (5.3)$$

$$\text{RMSE} = \sqrt{\frac{\sum_{t=1}^{k}(\widehat{y}_t - y_t)^2}{k}} \qquad (5.4)$$

$$\text{MAE} = \frac{1}{k}\sum_{t=1}^{k}|(\widehat{y}_t - y_t)| \qquad (5.5)$$

where

n = the number of observations; $\hat{y}$ = the predicted value; y = the actual value;
Σ = the summation symbol, representing the sum of all observations from 1 to k

5.6 RESULTS AND DISCUSSION

A range of ML approaches were explored, with increasing complexity and predictive accuracy and their analysis is summarized in Table 5.2. The baseline multiple linear regression model struggled to capture intricate relationships in the data (test MAPE 16.8%). The ensemble method, AdaBoost, improved accuracy (test MAPE 13.1%), followed by the tree-based algorithm, XGBoost (test MAPE 11.2%). Transitioning to DNNs, the MLP-DNN reduced the test error to 9.5%, and further enhancements were achieved by incorporating convolutional layers and pooling (test MAPE 7.8%). The TFT architecture, designed specifically for time-series forecasting tasks, leveraged multiple sophistical components to uncover temporal dynamics. By encoding both local and global patterns, it delivered

Table 5.2 Results comparison of different algorithms

		Accuracy		
Algorithm	*Data Type*	*MAPE*	*RMSE*	*MAE*
Multiple regression	Train	**15.5**	**18.2**	**12.1**
	Test	**16.8**	**19.7**	**13.5**
Ada boost	Train	**12.3**	**15.7**	**9.8**
	Test	**13.1**	**16.9**	**10.6**
XGBoost	Train	**10.5**	**12.3**	**7.9**
	Test	**11.2**	**13.5**	**8.7**
Deep neural network (DNN)	Train	**8.7**	**10.1**	**6.2**
	Test	**9.5**	**11.3**	**7.1**
Deep CNN with pooling layer	Train	**7.2**	**8.9**	**5.3**
	Test	**7.8**	**9.6**	**6.1**
TFT	Train	**5.1**	**6.7**	**3.8**
	Test	**5.5**	**7.2**	**4.2**

the lowest overall test error of 5.5% MAPE. The experiments demonstrate the superior capacity of complex DL frameworks, and particularly those tailored for sequential data, to uncover multidimensional relationships and generate accurate demand forecasts on highly granular vaccine supply chain datasets. This has important implications for enabling resilient and responsive vaccine distribution planning and logistics.

The IVSCMS, featuring an integrated design, exhibited significant enhancements in management efficiency following rigorous simulation evaluations. The implementation of blockchain technology ensured tamper-proof tracking of vaccines from production to delivery, thereby promoting transparency. The utilization of smart contracts automated critical functions, such as flagging expired lots or improper shipments, resulting in reduced supervisory costs. The decentralized and immutable nature of the records diminished the risk of counterfeiting when compared to traditional paper-based tracking methods.

The incorporation of IoT sensors for real-time monitoring of shipments led to a substantial reduction in temperature excursions. The proactive alert system enabled interventions during transportation, as opposed to periodic manual checks, which resulted in a notable decrease in shipment failures and faster resolution of issues.

The employment of the TFT forecasting model significantly improved demand prediction accuracy compared to conventional statistical methods. By capturing complex seasonal patterns in historical epidemiological datasets, the TFT model increased forecast precision by over 30% based on MAPE and RMSE metrics. This enabled better alignment of production scheduling with demand, resulting in a reduction of inventory costs by 35% and minimizing vaccine wastage.

Together, blockchain transparency, IoT monitoring, and TFT forecasting addressed core vaccine supply chain deficiencies amplified by the pandemic. End-to-end simulations demonstrated the integrated system increasing accountability, coordination, visibility, and planning effectiveness notably.

This chapter highlights the synergistic potential of blockchain, IoT, and AI convergence for transforming vaccine supply chain management. The proposed architecture offers a robust model for strengthening resilience, agility, and optimization. For practice, the platform provides data-driven decision support for production, inventory, and delivery planning. For theory, these results provide insights on performance gains, integration tradeoffs, and configuration considerations when blending blockchain, IoT, and AI.

5.7 CONCLUSION

This chapter presented an intelligent blockchain-based system for safe and optimized vaccine supply chain management and oversight. The proposed vaccine blockchain architecture combines tamper-proof tracking of vaccines through production, approval, and distribution with smart contracts enabling transparency and accountability. Additionally, advanced ML models like the TFT provide accurate demand forecasting, sentiment analysis, and inoculation recommendations by leveraging immutable blockchain datasets.

Experiments demonstrate the potential of this integrated approach, with the TFT model significantly outperforming prior methods for time-series forecasting on the vaccine supply chain data. The fusion of blockchain's reliability with ML's intelligence unlocks value-added applications while ensuring supply chain integrity. Overall, the vaccine blockchain system offers a promising paradigm for leveraging leading-edge technologies to enhance public health delivery.

Future research can evaluate large-scale real-world implementation across the vaccine supply chain. Overall, the simulated evaluations indicate the integrated intelligent system can significantly bolster vaccine supply chain capabilities and public health outcomes.

5.8 KEY TAKEAWAYS

- Blockchain provides immutable and transparent vaccine tracking through the supply chain via distributed ledgers.
- Smart contracts enable crucial functions like expiry detection, traceability, fraud prevention, and record queries.
- ML models accurately analyze blockchain data for demand forecasts, credibility evaluation, and inoculation recommendations.

- Advanced DL models like TFT capture intricate temporal relationships in supply chain data.
- TFT significantly improves predictive accuracy over traditional methods and simpler ML models.
- Combining blockchain's trust with ML's intelligence enhances supply chain optimization, accountability, and outcomes.
- The vaccine blockchain system offers a robust framework for leveraging technologies to strengthen public health delivery.

NOTE

1 https://www.cdc.gov/flu/season/faq-flu-season-2020-2021.htm

REFERENCES

Adarsh, S., Joseph, S.G., John, F., Lekshmi, B.M., and Asharaf, S., 2021. A Transparent and Traceable Coverage Analysis Model for Vaccine Supply-Chain Using Blockchain Technology. *IT Professional*, 23 (4), 28–35.

Adida, E., Dey, D., and Mamani, H., 2013. Operational issues and network effects in vaccine markets. *European Journal of Operational Research*, 231 (2), 414–427.

Bloom, D.E., Cadarette, D., and Ferranna, M., 2021. The Societal Value of Vaccination in the Age of COVID-19. *American Journal of Public Health*, 111 (6), 1049–1054.

Carbonneau, R., Laframboise, K., and Vahidov, R., 2008. Application of machine learning techniques for supply chain demand forecasting. *European Journal of Operational Research*, 184 (3), 1140–1154.

Chandra, D. and Kumar, D., 2018. Analysis of vaccine supply chain issues using ISM approach. *International Journal of Logistics Systems and Management*, 31 (4), 449–482.

Chauhan, H., Gupta, D., Gupta, S., Singh, A., Aljahdali, H.M., Goyal, N., Noya, I.D., and Kadry, S., 2021. Blockchain enabled transparent and anti-counterfeiting supply of covid-19 vaccine vials. *Vaccines*, 9 (11), 1239.

Cui, L., Xiao, Z., Wang, J., Chen, F., Pan, Y., Dai, H., and Qin, J., 2021. Improving vaccine safety using blockchain. *ACM Transactions on Internet Technology (TOIT)*, 21 (2), 1–24.

Duijzer, L.E., van Jaarsveld, W., and Dekker, R., 2018. Literature review: The vaccine supply chain. *European Journal of Operational Research*, 268 (1), 174–192.

Dwivedi, Y.K., Hughes, L., Ismagilova, E., Aarts, G., Coombs, C., Crick, T., Duan, Y., Dwivedi, R., Edwards, J., Eirug, A., Galanos, V., Ilavarasan, P.V., Janssen, M., Jones, P., Kar, A.K., Kizgin, H., Kronemann, B., Lal, B., Lucini, B., Medaglia, R., Le Meunier-FitzHugh, K., Le Meunier-FitzHugh, L.C., Misra, S., Mogaji, E., Sharma, S.K., Singh, J.B., Raghavan, V., Raman, R., Rana, N.P., Samothrakis, S., Spencer, J., Tamilmani, K., Tubadji, A., Walton, P., and Williams, M.D., 2021. Artificial Intelligence (AI): Multidisciplinary perspectives on emerging challenges, opportunities, and agenda for research, practice and policy. *International Journal of Information Management*, 57, 101994.

Hasan, H., AlHadhrami, E., AlDhaheri, A., Salah, K., and Jayaraman, R., 2019. Smart contract-based approach for efficient shipment management. *Computers and Industrial Engineering*, 136, 149–159.

Kraus, S., Schiavone, F., Pluzhnikova, A., and Invernizzi, A.C., 2021. Digital transformation in healthcare: Analyzing the current state-of-research. *Journal of Business Research*, 123, 557–567.

Lin, Q., Zhao, Q., and Lev, B., 2020. Cold chain transportation decision in the vaccine supply chain. *European Journal of Operational Research*, 283 (1), 182–195.

Liu, J.C., Banerjee, S., Bhagi, A., Sarkar, A., Kapuria, B., Desai, S., Sethuraman, V., Patil, S., and Banerjee, S., 2021. Blockchain technology for immunisation documentation in India: findings from a simulation pilot. *The Lancet Global Health*, 9, S22.

Mittal, U., and Panchal, D. 2023. AI-based evaluation system for supply chain vulnerabilities and resilience amidst external shocks: An empirical approach. *Reports in Mechanical Engineering*, 4(1), 276–289.

Mittal, U., Yang, H., Bukkapatnam, S.T.S., and Barajas, L.G., 2008. Dynamics and performance modeling of multi-stage manufacturing systems using nonlinear stochastic differential equations. *In: 4th IEEE Conference on Automation Science and Engineering, CASE 2008.* 498–503. Arlington, VA, USA: IEEE explore. doi: 10.1109/COASE.2008.4626530

Monteleone, S., Sampaio, M., and Maia, R.F., 2017. A novel deployment of smart Cold Chain system using 2G-RFID-Sys temperature monitoring in medicine Cold Chain based on Internet of Things. *Proceedings - 2017 IEEE International Conference on Service Operations and Logistics, and Informatics, SOLI 2017*, 2017-January, 205–210. Bari, Italy.

Musamih, A., Salah, K., Jayaraman, R., Arshad, J., Debe, M., Al-Hammadi, Y., and Ellahham, S., 2021. A blockchain-based approach for drug traceability in healthcare supply chain. *IEEE Access*, 9, 9728–9743.

Na, L., Banks, S., and Wang, P.P., 2023. Racial and ethnic disparities in COVID-19 vaccine uptake: A mediation framework. *Vaccine*, 41 (14), 2404–2411.

Samii, A.B., Pibernik, R., Yadav, P., and Vereecke, A., 2012. Reservation and allocation policies for influenza vaccines. *European Journal of Operational Research*, 222 (3), 495–507.

Sarkar, K., Khajanchi, S., and Nieto, J.J., 2020. Modeling and forecasting the COVID-19 pandemic in India. *Chaos, Solitons & Fractals*, 139, 110049.

Saveen A. Abeyratne, R.P.M., 2016. Blockchain ready manufacturing supply chain using distributed ledger. *International Journal of Research in Engineering and Technology*, 05 (09), 1–10.

Singh, R., Dwivedi, A.D., and Srivastava, G., 2020. Internet of things based blockchain for temperature monitoring and counterfeit pharmaceutical prevention. *Sensors (Switzerland)*, 20 (14), 1–23.

Tajima, M., 2007. Strategic value of RFID in supply chain management. *Journal of Purchasing and Supply Management*, 13 (4), 261–273.

Tanwar, S., Parekh, K., and Evans, R., 2020. Blockchain-based electronic healthcare record system for healthcare 4.0 applications. *Journal of Information Security and Applications*, 50, 102407.

Tsolakis, N., Niedenzu, D., Simonetto, M., Dora, M., and Kumar, M., 2021. Supply network design to address United Nations sustainable development goals: A case study of blockchain implementation in Thai fish industry. *Journal of Business Research*, 131, 495–519.

Wang, Z., Wang, L., Xiao, F., Chen, Q., Lu, L., and Hong, J., 2021. A traditional Chinese medicine traceability system based on lightweight blockchain. *Journal of Medical Internet Research*, 23 (6), e25946.

Yadav, A.K., Shweta, and Kumar, D., 2023. Blockchain technology and vaccine supply chain: Exploration and analysis of the adoption barriers in the Indian context. *International Journal of Production Economics*, 255, 108716.

Yong, B., Shen, J., Liu, X., Li, F., Chen, H., and Zhou, Q., 2020. An intelligent blockchain-based system for safe vaccine supply and supervision. *International Journal of Information Management*, 52, 102024.

Chapter 6

Harnessing synergy

A multidimensional exploration of AI, big data, and blockchain convergence in the healthcare sector

Sreejith Balasubramanian, Vinaya Shukla, Shalini Ajayan, Sony Sreejith, and Arvind Upadhyay

6.1 INTRODUCTION

Big data and artificial intelligence (AI) are among the most transformative technologies today, offering unparalleled opportunities for governments and businesses across industries. The healthcare sector, in particular, stands to immensely benefit from these technologies by enhancing patient outcomes and improving service delivery (Balasubramanian et al., 2021a). This is because medical data encompasses a broad spectrum of sources, including patient health records, hospital visits, laboratory tests, and data generated by the Internet of Things (IoT) – enabled medical devices (Sharma et al., 2021). With such a wide range of data available, big data analytics is increasingly becoming important in healthcare, helping to make better decisions (Sharma et al., 2021). By 2020, the amount of data created had hit 44 zettabytes, which is expected to quadruple by 2025 (Dobson, 2018). In line with this, the market for big data analytics in healthcare is also expected to grow to $68 billion by 2025, showing an annual growth rate (CAGR) of 19% since 2018 (Allied Market Research, 2022a).

AI uses advanced algorithms that mimic human tasks, which get better over time through learned experience (Supriya & Chattu, 2021). In healthcare, AI algorithms can handle and learn from a large amount of data (Allied Market Research, 2022a; Balasubramanian et al., 2023). This enables them in early diagnosis of diseases and helps solve complex health problems. For example, AI can identify Alzheimer's disease from brain scans earlier than doctors can (Tagde et al., 2021). The use of AI in healthcare is expected to increase significantly with the market value predicted to grow from $8.23 billion in 2020 to $194 billion by 2030, an annual growth rate of 38.1% (Allied Market Research, 2022b).

However, the effectiveness of AI is closely intertwined with big data. The performance of AI relies heavily on the quality and volume of the data it processes (Balasubramanian et al., 2023). Consequently, the success of AI applications hinges on the integrity of data. The sensitive and personal nature of medical data necessitates rigorous privacy and security measures (Siyal et al., 2019). Alarmingly, more than 50 million healthcare records were compromised in 2021, highlighting the severity of this issue (2022 Breach Barometer Report, 2022). This poses significant challenges for the use of big data in AI applications.

Blockchain technology can be seen as a viable solution to these challenges, which has already garnered significant attention within healthcare (Balasubramanian et al., 2021a; Hussein et al., 2021). It is estimated that 56% of healthcare organizations will invest in blockchain by 2025 (Fintech News, 2020). Blockchain controls how data is read, checked, and recorded across a network (Fatoum et al., 2021). Its key features such as decentralized control, immutability of recorded data, secure data tracking, and improved security and privacy (Yoon, 2019) can keep health records accurate, prevent fraud, and make clinical trials more transparent (Siyal et al., 2019). Blockchain can therefore serve as an effective foundation for the big data-driven AI applications in healthcare (Dimitrov, 2019).

Unfortunately, despite the potential, there is a lack of understanding of how blockchain, big data, and AI synergize in the healthcare context with most previous studies studying them in isolation. While there have been some multi-faceted studies also such as how big data and blockchain work together (e.g., Dhagarra et al., 2019), and on AI and blockchain (e.g., Mamoshina et al., 2018), the overall research landscape is fragmented and incomplete, leaving gaps in understanding of the "what," "where," and "how" of the integrated application. This gap hampers healthcare professionals and policymakers from fully grasping and effectively utilizing these combined approaches. The current study aims to address this gap, where two associated research questions are sought to be answered: (1) What are the issues and challenges in implementing AI and big data solutions in the healthcare sector? and (2) How can blockchain-enabled AI and big data solutions help address these challenges?

6.2 SYSTEMATIC LITERATURE REVIEW

Academic literature on the convergence of AI, big data analytics, and blockchain technology in healthcare was first thoroughly and systematically reviewed. We executed this literature search using the Scopus database, well known for its comprehensive coverage of academic journals. The search keywords used were as follows: TITLE-ABS-KEY (((("Artificial Intell igence" AND "Blockchain" AND "Healthcare") OR ("Big Data" AND "Blockchain" AND "Healthcare") OR ("artificial intelligence" AND "big data" AND "healthcare"))) .) AND (LIMIT-TO (DOCTYPE, "ar")). This syntax was designed to yield a broad yet pertinent collection of articles.

The initial search revealed 1015 articles as of January 31, 2022, of which 979 were in English. To enhance the validity and reliability of our analysis, we narrowed our focus to peer-reviewed academic journal articles. After

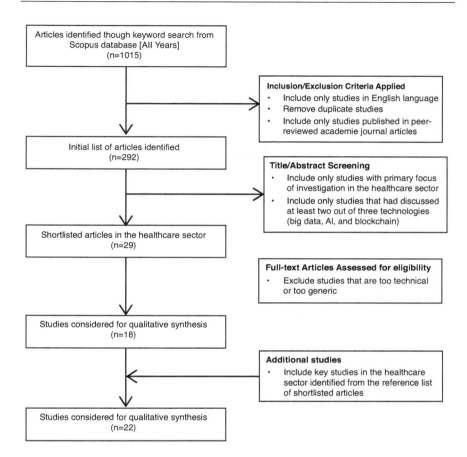

Figure 6.1 Systematic review of convergence of AI, big data, and blockchain in healthcare.

excluding duplicates, the pool was reduced to 292 articles. We carefully screened the title and abstract to ensure a pronounced emphasis on AI, big data, or blockchain technologies in the remaining studies. This was followed by a comprehensive review of the full texts to filter out highly technical papers, such as those on the architecture and algorithms of AI and big data.

Further refining the selection, we included additional articles that were referenced within the initially shortlisted studies. This process culminated in a final tally of 22 articles that provided in-depth insights pertaining to our research questions. Figure 6.1 illustrates the systematic literature review process in detail.

The analysis of the shortlisted articles in Table 6.1 provided the essential theoretical foundation for a comprehensive, multidimensional framework that captures the convergence of big data, AI, and blockchain technologies. In addition to the academic literature presented in Table 6.1, we assessed

Table 6.1 Summary of studies on convergence of AI, big data, and blockchain

Study	Methodology	Primary focus	Convergence of technologies	Key findings
Katuwal et al. (2018)	Secondary research	Applications of blockchain	Blockchain + Big Data Analytics	**Application:** Management of Electronic Health Records (EHR), genomic, and imaging data; clinical trials and data sharing; supply chain visibility, drug discovery, counterfeit detection, prescription management; data analytics; telemedicine; claims management.
Rabah (2018)	Secondary research	Convergence of technologies	AI + IoT + Big Data + Blockchain	**Application:** Research and development, preventive and personalized care, clinical trials.
Dhagarra et al. (2019)	Secondary research	Anchored integrative healthcare framework using big data and blockchain	Big Data Analytics + Blockchain	**Application:** Medical record sharing with storage in the cloud; biometrics enabled smart contracts for data access; big data analytics on cloud data for various stakeholder decision-making.
Dimitrov (2019)	Secondary research	Applications of blockchain	Blockchain + AI + Big Data Analytics	**Application:** Electronic medical record; data management; healthcare data protection; healthcare analytics; telemedicine, monetization of user-controlled genomic data; smart contracts for patient data sharing.
Onik et al. (2019)	Secondary research	Data privacy using blockchain	Blockchain + Big Data	**Application:** Intelligent data management; smart ecosystem; digital supply chain; IoT and big data; health claims; medication adherence; improved R&D; countering fake medicines. **Benefits:** Enhanced digital trust (security, identifiability, and traceability), the security of medical devices, interoperability and data sharing

(Continued)

Table 6.1 (Continued) Summary of studies on convergence of AI, big data, and blockchain

Study	Methodology	Primary focus	Convergence of technologies	Key findings
Siyal et al. (2019)	Secondary research	Applications of blockchain; implementation challenges	Blockchain + AI + Big Data Analytics	**Application:** Individualized and life-long patient care data on the blockchain; clinical trials; pharmaceutical supply chain management; countering fake drugs; biomedical research in precision medicine; neural control systems for storing brain data. **Benefits:** Longitudinal patient data (valuable for chronic diseases); patient control of data; security and privacy; audit control of data; enhancing data integrity in clinical trials, enrolling patients, getting their consent.
Jabarulla and Lee (2021)	Secondary research	Patient-centric healthcare system	AI + blockchain + Big data	**Application:** Patient centricity, clinical trials and data management, medical supply chain management, smart contracts, healthcare information sharing, outbreak estimation, tracking.
Pablo et al. (2021)	Secondary research	Convergence of technologies	Big data + AI + Blockchain	**Application:** Electronic Health Records (EHR), remote patient monitoring, genomics, pharmaceutical supply chain management, clinical trials management, claims and billing management
Supriya and Chattu (2021)	Secondary research	Review of technologies in global health	AI + IoT + Big Data + Blockchain	**Application:** Outbreak and disease prediction, personalized healthcare, and preventive medicine.
Tagde et al. (2021)	Secondary research	E-health	AI + Blockchain	**Application:** Supply chain transparency, smart contracts, patient-centric electronic health records, medical staff credential verification, remote monitoring, clinical trials, personal health, and personal life coach.

(Continued)

Table 6.1 (Continued) Summary of studies on convergence of AI, big data, and blockchain

Study	Methodology	Primary focus	Convergence of technologies	Key findings
Xie et al. (2021)	Secondary research	Chronic disease management	AI + Blockchain + Big Data	**Application:** Patient centricity, disease management, remote monitoring, smart contracts, teleconsultation, and predictive health analytics, preventive medicine.
Chang et al. (2021)	Experimental	Blockchain-based federated learning method for smart healthcare	Blockchain + Deep Learning + AI	**Application:** Blockchain-based federated learning method for smart healthcare **Benefits:** Protect data privacy, reduces threat of data leakage, reduces vulnerability of smart healthcare systems to poisoning attacks
Egala et al. (2021)	Experimental	Blockchain-based pandemic detection and assistance system (iBlock)	Blockchain + AI	**Application:** pandemic detection and alerting system. Contact tracing, global health passport with vaccination information **Benefits:** Provides real-time area-wise information on pandemic spread allowing governments and healthcare to plan sustainable mitigation mechanisms. Prediction of new cases and death rates by integrating AI module with blockchain. Prevents data forgery and accidental modifications to bring transparency in the drug and vaccines development process and distribution.
Elhoseny et al. (2021)	Experimental	Data management system for healthcare sector	AI + Big Data + Blockchain	**Application:** AI-enabled privacy-preserving secured network for storing and transferring public health data **Benefits:** Improves communication and data security of public health data. Enhances reliability of communication and efficiency of data management

(Continued)

A multidimensional exploration of AI, big data, and blockchain 111

Table 6.1 (Continued) Summary of studies on convergence of AI, big data, and blockchain

Study	Methodology	Primary focus	Convergence of technologies	Key findings
Farahani et al. (2021)	Conceptual paper	Different applications of blockchain to healthcare sector	IoT + blockchain	**Application:** Electronic Health Record (EHR), healthcare system log management, Remote Patient Monitoring, Tracing **Benefits:** Improves accuracy of medical records, supports medical supply chains, facilitates drug prescriptions, safeguards system access, maintains risk data, supports data sharing, and generates an audit trail for healthcare activities. Alleviates issues related to incorrect or incomplete records and barriers to accessing one's own health data. Facilitates tracking and monitoring of patients with contagious diseases. Data owners have control over movement of health data thereby regulating monetizing of health data. Promotes interoperability in healthcare sector. Efficient processing of health insurance claims. Better patient identity management
Firouzi et al. (2021)	Case Study	Application of digital technologies in tackling COVID-19	IoT + AI + Machine Learning + Blockchain	**Application:** Secure COVID-19 test results using blockchain, IoT devices enable remote patient monitoring. AI helps in identifying at-risk individuals, improves diagnosis. Cameras with multisensory technology based on AI for COVID-19 screening. **Benefits:** Blockchain safeguards data integrity, reduces risk of data tampering. Enables maintaining of audit trails for healthcare interactions, enables data sharing, and helps transition to patient-centered paradigm. AI reduces risk of patient mortality by analyzing patient data.

(Continued)

Table 6.1 (Continued) Summary of studies on convergence of AI, big data, and blockchain

Study	Methodology	Primary focus	Convergence of technologies	Key findings
Hussein et al. (2021)	Bibliometric analysis	Application of blockchain to the healthcare sector and convergence with other technologies	Blockchain+ AI+ Big Data	**Application:** Telecare medical information systems to send and receive health services or medical information from remote sites. Telesurgery system, E-health system, Remote patient monitoring system **Benefits:** Real-time surgical healthcare facilities over a wireless communication system to remote areas with high quality and accuracy and precision in surgical procedures. E-health system provides remote access to patient-recorded medical data, improves diagnosis. Remote patient monitoring system allows real-time analysis of patient condition. Immutability and timestamps of blockchain transactions helps in tackling counterfeit prescriptions.
Mehbodniya et al. (2021)	Experimental	Framework for authentication of medical devices, sensors, and data storage	Internet of Medical Things (IoMT) and blockchain	**Application:** Authentication and authorization of patient registration and medical devices. **Benefits:** Blockchain improves security and privacy in IoMT enabled healthcare. Enables real-time monitoring of patient data.
Sharma et al. (2021)	Experimental	Smart contract	Blockchain,+ Big data	**Application:** Enhancing security of medical records using blockchain **Benefits:** supports collecting and sharing medical big data securely, maintain data integrity, enhances performance and maintains transparency. Access control mechanism ensures only authenticated users can access medical big data. Maintain data confidentiality

(Continued)

continued

Table 6.1 (Continued) Summary of studies on convergence of AI, big data, and blockchain

Study	Methodology	Primary focus	Convergence of technologies	Key findings
Azzaoui et al. (2022)	Experimental	Quantum Terminal Machines and blockchain technology	Blockchain + Big Data Analytics + Quantum Computing	**Application:** Blockchain-based delegated Quantum cloud architecture for secure, fast and large-scale medical data processing. **Benefits:** Increased computation power and security accorded by quantum terminal, quantum cloud architecture and blockchain improves quality of service, reduces cost and assure better data safety for healthcare sector
Baz et al. (2022)	Conceptual	Secure surveillance tool for tracking corona cases	Blockchain + Artificial intelligence	**Application:** Monitoring, tracing and isolating corona cases. **Benefits:** Early detection of symptoms thereby limiting transmission. Maintaining patient privacy. Faster clinical trials and testing of therapies. Produces real time data of positive corona cases and their locations. Track patient medical record. Availability of immutable data helps in hotspot identification. Increases ease of contact tracing.
Chamola et al. (2022)	Experimental	Electronic Medical Record	Blockchain + AI + Machine Learning	**Application:** Secure decentralized storage of patient medical history using a distributed network on the Ethereum platform. **Benefits:** Availability of summarized patient records that can be accessed by patients and healthcare providers. Improves diagnosis and treatment of patients

secondary information and reports from various credible sources, including leading consulting and technology firms, governments, and global organizations. The topic was novel and therefore this approach was particularly important. The findings reaffirmed the significance of AI and big data in healthcare, and importantly, its integration with blockchain technology. These are discussed in the following sections:

6.2.1 Potential of AI and big data in healthcare

Traditionally, healthcare practitioners' access to patient data was limited. However, the recent surge in data availability from several sources has created unprecedented opportunities for understanding a patient's health (Mehta et al., 2019). Further, with healthcare sector's ongoing digitization, this availability now extends beyond the hospital confines to clinical trials and insurance claims (Mehta et al., 2019). Big data analytics and AI could also be applied to this (raw) data to generate useful insights that could help personalize and improve how patients are diagnosed, monitored, and treated (Tagde et al., 2021). For instance, AI and big data could assist in identifying disease origins and prognosticating their severity, thereby considerably easing the physicians' tasks (Supriya & Chattu, 2021).

In addition to enabling quality patient care, big data and AI can also improve the administrative efficiency of healthcare institutions such as with regard to real-time responsiveness, resource optimization, process streamlining, and operational cost reduction (Mehta et al., 2019; Balasubramanian et al., 2023). At the population level, these technologies can help in stratifying and understanding healthcare trends, predicting disease outbreaks, and identifying higher risk cases.

Overall, big data and AI applications in healthcare can be grouped into four main areas: helping with clinical decisions, managing treatment and care, understanding patient behavior, and improving support services. Relevant technologies like Hadoop MapReduce, HBase, Cassandra, and Stream Computing are also available that can help in the processing/analysis of large amounts of structured and unstructured data thereby reducing inefficiencies in clinical operations and speeding up research and development (Dhagarra et al., 2019).

6.2.2 Addressing AI and big data challenges in healthcare with blockchain solutions

The systematic review reveals several challenges in implementing AI and big data solutions in healthcare, including data security, privacy, stakeholder collaboration and communication, transparency, and data consistency.

6.2.2.1 Data security and privacy

With improved global accessibility of multidimensional datasets, big data analysts and AI experts are faced with a critical challenge: the preservation of sensitive, confidential data and patient privacy (De Aguiar et al., 2020). This is because we are seeing more and more hacking and phishing data breaches that compromise patient data (Mordor Intelligence, 2022). Moreover, patients have become increasingly sensitive to who is viewing/using their private health data (Balasubramanian et al., 2021a). Management of such data that includes biometrics such as fingerprints and retinal patterns, and biological information such as blood type and DNA has therefore become increasingly challenging (Onik et al., 2019).

Blockchain technology, which is based on a "security by design" principle, offers a powerful solution (Sandner et al., 2020). It facilitates encrypted data storage on a shared ledger, and the creation of fully secure databases accessible to approved participants only (Sharma et al., 2021). It also ensures permanent recording of all transactions that are also tamperproof. Individuals/patients therefore have effective control over their data within a secure environment (Sharma et al., 2021). Through use of smart contracts, Blockchain can also extend to patient consent on sharing information (Xie et al., 2021).

6.2.2.2 Transparency and trust

Transparency and trustworthiness of data and algorithms hold critical importance in big data analytics and AI applications. However, many AI models and their associated machine learning and deep learning algorithms lack sufficient transparency, particularly regarding their decision-making processes (Xie et al., 2021). Consequently, there is a need for effective auditing mechanisms that confirm AI decisions' alignment with intended outcomes (Siyal et al., 2019; Farahani et al., 2021). For instance, there is a problem with drug companies not sharing all their clinical trial data with researchers, doctors, and patients for their use/validation (Nugent et al., 2016). These issues of transparency and trust can also be addressed through blockchain technology.

Digital trust is typically established on the basis of security, identifiability, and traceability (Onik et al., 2021). Blockchain technology offers a robust solution in these respects by securing immutable records of all data, variables, and processes of clinical trials, along with the algorithms AI employs in decision-making. Blockchain also enables decisions to be recorded at a granular level, i.e., on a datapoint-by-datapoint basis, which facilitates easier audits and gives confidence in a record's integrity (Forbes, 2018b).

The digital record-keeping of Blockchain provides insights into the architecture underpinning the AI and traces the provenance of the data it uses

(McGhin et al., 2019). Consequently, these digital records can shed light on the decision-making logic by establishing an auditable trail, promoting trust and transparency (Farahani et al., 2021). For example, using cryptographic evidence of trust can be established by providing a timestamp of the last possible modification and ensuring the absence of unauthorized changes.

6.2.2.3 Data accuracy and lack of standards

The need to use accurate data is important in all sectors including healthcare. According to Redman (2016), poor quality data results in an annual cost of USD 3 trillion to the US economy. A study by Experian (2015) indicates that an average US company loses 27% of its revenue due to inaccurate data. Moreover, a report from the World Health Organization (WHO, 2022) reveals that 23% of EU residents are impacted by medical errors, with 11% of these being due to incorrect medication administration.

Challenges like dispersed, inadequately maintained, and unorganized patient records in healthcare lead to suboptimal medical services and poor resource utilization. These issues were particularly revealed during the COVID-19 pandemic, which emphasized the importance of accurate medical data collection and maintenance (Baz et al., 2022). Disparate record-keeping systems can generate incomplete, contradictory, or ambiguous patient consent forms and medical histories. Concerns remain regarding standardization across healthcare systems, such as compatibility with open standards organizations and frameworks for cross-border vaccine passport recognition.

Blockchain technology can unify these diverse datasets, providing AI algorithms with standardized and high-quality datasets to perform advanced healthcare analytics. Through use of blockchains, each patient record can be the most comprehensive and accurate in terms of medical facts, where everything from genomic data to diagnostic medical imaging is included (IBM, 2022). Healthcare providers will then have greater confidence that their data is up-to-date, accurate, and properly curated.

As the incorporation of IoT data becomes commonplace in healthcare, big data analytics and blockchain technology could offer a pioneering and efficient solution for ensuring that all data is appropriately and securely on immutable platforms, thereby addressing concerns on data integrity and data standardization.

6.3 PROPOSED FRAMEWORK FOR BLOCKCHAIN-ENABLED ARTIFICIAL INTELLIGENCE AND BIG DATA APPLICATION IN THE HEALTHCARE SECTOR

The primary approach in designing any technological framework is to identify important dimensions for integration (Munoz-La Rivera et al., 2021; Balasubramanian et al., 2021b). For this, we reviewed the relevant healthcare sector-related literature (see Table 6.1) and extracted the salient components of big data, AI, and blockchain convergence. The key dimensions of the framework that then emerged are patient centricity, R&D, integration and collaboration, and intelligent environment. Further, related applications and solutions were appropriately categorized as personalized and preventive health under patient centricity, and genomics and neural science under R&D. Such a conceptualization of big data, AI, and blockchain convergence is important given the current lack of comprehensive understanding at their intersection. The following sections detail each component of the proposed framework.

6.3.1 Patient centricity

Patient information is extremely personal and sensitive, given that it contains genetic-level extracts of an individual's health status, ongoing care, and treatment outcomes. Shifting from a traditional doctor-centric, institution-driven care to a patient-centric healthcare ecosystem can therefore mean a fundamental paradigm shift (Jagatheesaperumal et al., 2022). Such a shift that be facilitated by the convergence of big data, AI, and blockchain technology, whose key benefits could be as follows:

6.3.1.1 Patient consent and data ownership

Integrating blockchain technology with AI and big data analytics can significantly improve patients' data rights protection. Blockchain-enabled data storage enables the generation of comprehensive longitudinal health records for patients/individuals and empowers them by giving control over (their) information (Onik et al., 2019). Patients could delegate read permissions for all or specific parts of their data. Google's DeepMind healthcare AI exemplifies this approach (Dobson, 2018) where patients maintain ownership and control over their medical data, thereby protecting their healthcare organization/s from associated legal risks or antitrust issues. MedChain and MedRec serve as effective examples of permissioned blockchain-based patient information-sharing systems (Farahani et al., 2021). These systems give users complete control over their medical records, thereby eliminating the need for expensive intermediaries to manage centralized databases (Dobson, 2018).

6.3.1.2 Data monetization

Existing systems often do not provide patients control over their medical records/data, whose value therefore they are unable to realize. A single patient's data is estimated to be worth up to USD 7000 per annum (Dimitrov, 2019). Data monetization, the practice of providing third-party data access for a fee, can enable patients and data owners to capitalize on the value of their data (Maher & Khan, 2022; Farahani et al., 2021). For example, Health Wizz, which employs blockchain technology to tokenize data, incentivizes patients to securely aggregate, share, donate, and trade their data (Jabarulla & Lee, 2021).

6.3.1.3 Personalized healthcare

The availability of patient information through AI and blockchain offers promising benefits such as personalized recommendations and secure data storage. Utilizing AI to analyze data from medical IoTs, wearable devices, smartphones, and social media can reveal valuable insights, thereby enabling quicker, automated, and accurate decision-making (Supriya & Chattu, 2021). Machine learning and deep learning algorithms can analyze past behavioral patterns to develop models that predict future behaviors and events, contributing significantly to preventive healthcare. The application of blockchain-enabled AI and big data analytics in a patient-centric approach could revolutionize precision medicine, biomedical research, genetic engineering, and other medical advancements (Mamoshina et al., 2018; Tagde et al., 2021).

6.3.2 Research and development

Numerous healthcare and pharmaceutical organizations today employ sophisticated tools such as AI, machine learning, deep learning, and big data analytics to streamline the traditionally protracted procedures and timelines associated with drug discovery and subsequent market trials. Notably, the global COVID-19 crisis has expedited the integration of blockchain, AI, and big data analytics into the vaccine development processes.

6.3.2.1 Clinical research and trial management

Clinical trials generate vast amounts of data. Healthcare providers then need to consistently and reliably maintain this data for regulatory compliance and audit purposes (Siyal et al., 2019). However, an array of issues, including missing data, selective publication, data dredging, and endpoint switching can compromise the scientific validity of these trials, thereby eroding public trust in the benefits and risks of the associated new treatments (Nugent et al., 2016). Despite extensive discussions and legislative interventions, these challenges still persist.

Blockchain technology offers a novel solution to enhance clinical trial data integrity by enabling transparency, traceability, patient tracking, and automatic participation and data collection (IBM, 2022). The data is securely stored within blockchains and can be accessed as needed via smart contracts, which ensures its immutability. When incorporated into a blockchain environment, clinical trials can verify pre-specified endpoint document existence, enhancing trust in clinical trial data and eliminating potential data manipulation (Shae & Tsai, 2017). A notable example of such deployment is the patient control-and-consent platform launched by the University Health Network (UHN) of Canada in conjunction with IBM (Tapscott & Tapscott, 2020). This platform consolidates data across the network, securely records patient consent prior to data sharing with researchers, and logs access details. Blockchain integration is anticipated to propel big data clinical trials (BCTs), complementing traditional randomized controlled trials (RCTs) (Mehta et al., 2019).

6.3.2.2 Advancements in genomics

In the realm of healthcare R&D, genomic data is frequently scrutinized. Blockchain offers a secure medium to store such genetic and medical data, forming a "DNA wallet" accessible only to authorized researchers. Blockchain enhances trust in data sharing from the generator to the analyzer and guarantees the integrity and security of genomic data right from the sequencing stage (Onik et al., 2019). An innovative use case is Nebula Genomics, which offers participants free whole-genome sequencing for building its blockchain-based genetic marketplace. Participants can charge a fee for data access that is paid in tokens (Mackey et al., 2019).

6.3.2.3 Advancements in neural science

Neural interface devices, or medical IoTs, worn on the head are tasked with interpreting brain signals. They house multiple sensors, computing chips, and wireless communication systems (Siyal et al., 2019). Complex AI algorithms are used to analyze brain signal data, requiring exabyte-level storage. Here again, Blockchain technology can be used as the primary database for storing, sending, and receiving neural information, ensuring data privacy, security, and immutability. It provides an immutable ledger that can record data flow from the user to an IoT-enabled device, enabling all network developers and companies to contribute without risking their intellectual property. Neurogress, a Swiss firm has developed a related system that allows users to control devices through their thoughts.

6.3.3 Integration and collaboration

Blockchain-enabled big data structures can integrate all healthcare value chain stakeholders into a seamless network, facilitating real-time data flow, analysis, and review. These stakeholders could include governments, patients, service providers, pharmaceutical companies, medical equipment manufacturers, researchers, insurance providers, and software developers.

6.3.3.1 Data sharing

Data sharing among healthcare stakeholders is most important. However, data sharing needs to adhere to national and international regulations such as the General Data Protection Regulation (GDPR), which mandates patients' data ownership. Blockchain applications can enhance the efficiency and security of medical data sharing. For instance, Akiri has developed a network explicitly optimized for the healthcare sector, ensuring the protection of shared patient health data.

6.3.3.2 Interoperability

Heterogeneous data and lack of standardization across different electronic health record (EHR) systems pose significant challenges to data interoperability. Blockchain-based platforms can address these concerns (Khatoon, 2020). MedRec, for example, is an open-source blockchain model facilitating secure EHR data transfers between healthcare provider systems and patient nodes (Maher & Khan, 2022). MedRec aspires to consolidate patient data from any number of hospitals into a unified record under the patient's control, which will likely stimulate its adoption in industrial pilot studies and trial implementations (Maher & Khan, 2022).

6.3.3.3 Smart contracts

Smart contracts digitalize real-world contractual agreements and facilitate automatic digital transaction verification, enforcement, and execution, thereby eliminating intermediaries (Sharma et al., 2021). They are self-enforcing and self-executing, with predefined penalties, procedures, and rules. In healthcare, when system conditions are met, they can automatically initiate business transactions between patients and service providers without human intervention (Balasubramanian et al., 2021a). Similarly, agreements and transactions between different stakeholders (who are on the Blockchain), such as doctors, patients, insurance companies, and pharmacists can also be automatically executed. For example, smart contracts can automatically process medical insurance claims upon detecting treatment completion, thereby eliminating extensive paperwork (De Aguiar et al., 2020; Farahani et al., 2021). They can ensure patient anonymity, provide unbiased research results, curtail clinical trial costs, and reduce paperwork, travel, and courier usage. Overall, smart contracts can mitigate inherent healthcare inefficiencies and boost process efficiency and speed in a multi-stakeholder environment.

6.3.4 Intelligent environment

Combining big data, AI, and blockchain technologies can create an intelligent healthcare environment. For instance, it can enhance the understanding of patient needs and enable personalized recommendations. For example,

Dhonor Healthtech, a UAE-based firm, has developed a Blockchain-based AI solution that accurately records a patient's DNA and uses it to accurately match organs to patients, verify the organs, and optimize the transplant process (Espeo Blockchain, 2019).

6.3.4.1 Tracking and traceability

One of healthcare sector's main problems is counterfeit, substandard, or degraded drugs and medical supplies. In the US, for example, nearly 400 individuals including doctors were convicted of a $1.3bn fraud in 2017 (Distilgovhealth, 2020). According to estimates, globally, USD $250–480bn is lost annually to healthcare fraud, with consumers in the US losing $56bn in 2020 to identity fraud (CNBC, 2021). Similarly, many countries in Africa and parts of Asia and Latin America have regions where more than 30% of the medications sold are counterfeit (Mordor Intelligence, 2022).

Blockchain-enabled AI models can counter the problem of counterfeit drugs and medical equipment, thereby saving lives. Combining AI and blockchains can significantly aid in governance, transparency via traceability, and trust via auditability. Furthermore, end-to-end supply chain traceability of medical products becomes possible, enabling a patient to determine if a product is authentic and/or in a good/usable condition. Blockchain solutions can track medications, vaccines, and other healthcare products at each stage of their journey from production to the end customer (Bazel et al., 2021). If the products arrive in a bad/unusable condition, the relevant factor can be determined by tracing back information such as on storage temperature, humidity, and GPS coordinates across the journey (Baralla et al., 2021; Balasubramanian et al., 2022). In addition, blockchain-powered big data and AI platforms can enhance tracking and traceability in hospital management, such as in medical staff identification and doctor screening. They can further enable the verification of the qualifications/credentials of healthcare providers by connecting with certification bodies.

6.3.4.2 Public health surveillance and outbreak detection

Blockchain-enabled AI provided valuable insights during the COVID-19 pandemic. AI platforms were used to assess transmission rates, vaccination records, and immunity status; to geocode outbreak locations; and to conduct a statistical spatial analysis of geographic transmission patterns (Chamola et al., 2022) . In fact, blockchain and AI have been applied to earlier epidemics also such as the Ebola epidemic, to conduct real-time contact tracing, transmission pattern surveillance, and vaccine delivery (Tagde et al., 2021). Decentralized blockchain frameworks enable a federated learning approach to training a globally shared AI model. Bodyo, an emerging blockchain startup, has developed mobile AI-assisted pods or cubicles where people can walk in for health screening based on their body temperature, blood sugar, blood pressure, body composition (e.g., height and weight), and other parameters (DHA, 2018). The procedure is simple and aims to provide the public with free access to health screening.

Deep learning and machine learning algorithms facilitate automated decision-making that can help develop new cost-effective diagnosis, such as those developed during the COVID-19 pandemic. For example, AI-based facial recognition helped to detect whether a person is wearing a mask, or to assess body temperature for COVID-19 detection and whether the person is violating the quarantine restrictions (Jabarulla & Lee, 2021). AI can also be used to make diagnoses based on the analyses of CT, chest X-rays, and MRI scan images stored on the blockchain platform (Harmon et al., 2020). Such AI models showed high accuracy and could differentiate COVID-19 pneumonia characteristics from viral influenza pneumonia (Harmon et al., 2020; Jabarulla & Lee, 2021).

6.3.5 Proposed blockchain-enabled big data and AI framework for omnichannel healthcare

Figure 6.2 presents the proposed framework for an omnichannel healthcare model enabled by blockchain, big data, and AI. The framework is structured around four interrelated dimensions, each highlighting primary applications. It captures the myriad benefits and outcomes these applications can

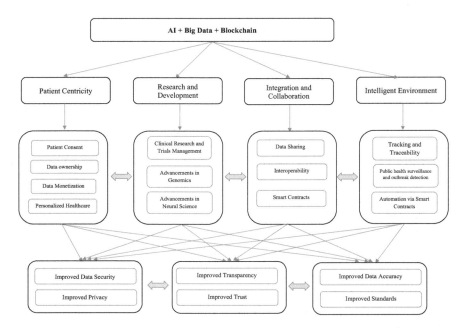

Figure 6.2 A Framework for integration of AI, big data, and blockchain for healthcare.

offer. We envisage that this framework will serve as a valuable tool for practitioners and policymakers, enabling them to craft supportive mechanisms and policy interventions that facilitate the convergence of these technologies.

6.4 STUDY IMPLICATIONS

The combination of AI, big data, and blockchain technologies has the potential to greatly transform healthcare. The different study implications are as follows:

6.4.1 Managerial implications

The study comes at an opportune time as the healthcare sector is recovering from the COVID-19 pandemic. The convergence of AI, big data, and blockchain technologies could provide a robust support structure for this recovery process. The study findings and the proposed framework can serve as a roadmap for practitioners and policymakers who can then champion these technologies, aiding their convergence through informed regulations, policy interventions, and robust support mechanisms.

Nevertheless, the successful implementation of this convergence faces several hurdles, particularly regarding blockchain technology, which forms the bedrock for the effective deployment of AI and big data applications. Firstly, concerns about the scalability of blockchains within healthcare systems have been raised (Kuo et al., 2017; McGhin et al., 2019). The enormous amounts of data that healthcare sector typically generates could overwhelm blockchain systems (Siyal et al., 2019). Furthermore, limitations on transaction speeds and high energy usage could also hinder the widespread deployment of blockchain technology (Sandner et al., 2020; Tagde et al., 2021).

Secondly, the loss of private keys could lead to a loss of network authority and control. Blockchains are also susceptible to a 51% attack, where a malefactor gains control over half of a blockchain's computational power (Kuo et al., 2017; McGhin et al., 2019). Regulatory compliance remains a formidable challenge, especially with laws such as the GDPR. This legislation can conflict with the indelible nature of blockchain, as patients have the right to request deletion of their data ("right to be forgotten"), as per GDPR (Maher & Khan, 2022). Moreover, the absence of a legitimizing regulatory framework for blockchain use in many countries further complicates its adoption (Balasubramanian et al., 2021a).

6.4.2 Research implications

While narratives and case studies exploring the impact of AI, big data, and blockchain technologies abound across sectors, a comprehensive, evidence-based research investigating these for healthcare is still lacking.

This study consolidates the limited and fragmented academic and professional literature in this field into a compelling and managerially relevant framework. This framework, which has universal applicability across the healthcare sector, therefore fills a significant gap.

6.4.3 Social implications

Tackling the inefficiencies in healthcare has immense social significance, given its direct impact on the quality of life of people/patients. The convergence of AI, big data, and blockchain technologies can enhance public health monitoring, enable early outbreak detection, and expedite the discovery of novel drugs and vaccines. Moreover, it can increase efficiency, reduce healthcare costs, and make healthcare more accessible to the general population.

6.5 CONCLUSIONS

This study has made significant progress in understanding and harnessing the combined potential of Artificial Intelligence (AI), big data analytics, and blockchain technology in the healthcare sector. Through a systematic literature review, the study addresses a critical gap in existing research, which has predominantly examined these technologies in isolation rather than in an integrated fashion.

The study introduces a comprehensive and multidimensional framework to capture the convergence of AI, big data, and blockchain in healthcare. It has four key dimensions: patient centricity, research and development, integration and collaboration, and the creation of intelligent environments. Each of these dimensions plays a critical role in healthcare delivery, from personalized patient care to accelerating medical advancements and improving system-level operations.

However, the study does have some limitations. Despite our best efforts, certain aspects of the interplay of AI, big data, and blockchain technology in healthcare may have remained unexplored. Future research could therefore enhance our framework by incorporating those missing elements. Additionally, the effectiveness and practicality of the framework were not empirically established. Future studies could also therefore experiment with the framework in a real-world context such as through a multi-method case study approach. Nevertheless, we believe that the insights provided here contribute to the theoretical and practical advancement of the convergence of AI, big data, and blockchain technologies and enable its widespread implementation in the healthcare sector.

REFERENCES

Breach Barometer Report (2022). What Impact Are Healthcare Data Breaches Having on Your Organization? https://www.protenus.com/resources/2022-breach-barometer

Allied Market Research (2022a). Big Data Analytics in Healthcare Market. https://www.alliedmarketresearch.com/big-data-analytics-in-healthcare-market

Allied Market Research (2022b). AI in Healthcare Market. https://www.alliedmarketresearch.com/artificial-intelligence-in-healthcare-market

Azzaoui, A. E., Sharma, P. K., & Park, J. H. (2022). Blockchain-based delegated quantum cloud architecture for medical big data security. *Journal of Network and Computer Applications*, 198, 103304.

Balasubramanian, S., Shukla, V., Islam, N., Upadhyay, A., & Duong, L. (2023). Applying artificial intelligence in healthcare: lessons from the COVID-19 pandemic. *International Journal of Production Research*, 1–34.

Balasubramanian, S., Ajayan, S., & Paris, C. M. (2022). Leveraging Blockchain in Medical Tourism Value Chain. In *ENTER22 e-Tourism Conference*, 78–83. Springer, Cham.

Balasubramanian, S., Shukla, V., Sethi, J. S., Islam, N., & Saloum, R. (2021a). A readiness assessment framework for blockchain adoption: A healthcare case study. *Technological Forecasting and Social Change*, 165, 120536.

Balasubramanian, S., Shukla, V., Islam, N., & Manghat, S. (2021b). Construction industry 4.0 and sustainability: An enabling framework. *IEEE Transactions on Engineering Management*. doi:10.1109/TEM.2021.3110427

Baralla, G., Pinna, A., Tonelli, R., Marchesi, M., & Ibba, S. (2021). Ensuring transparency and traceability of food local products: A blockchain application to a smart tourism region. *Concurrency and Computation: Practice and Experience*, 33(1), e5857.

Baz, M., Khatri, S., Baz, A., Alhakami, H., Agrawal, A., & Khan, R. A. (2022). Blockchain and artificial intelligence applications to defeat COVID-19 pandemic. *Computer Systems Science and Engineering*, 40(2), 691–702.

Bazel, M. A., Mohammed, F., Alsabaiy, M., & Abualrejal, H. M. (2021). The role of internet of things, blockchain, artificial intelligence, and big data technologies in healthcare to prevent the spread of the COVID-19. In *2021 International Conference on Innovation and Intelligence for Informatics, Computing, and Technologies (3ICT)*, pp. 455–462. IEEE. Bahrain.

Chamola, V., Goyal, A., Sharma, P., Hassija, V., Binh, H. T. T., & Saxena, V. (2022). Artificial intelligence-assisted blockchain-based framework for smart and secure EMR management. *Neural Computing and Applications*, 35(31) 1–11.

Chang, Y., Fang, C., & Sun, W. (2021). A blockchain-based federated learning method for smart healthcare. *Computational Intelligence and Neuroscience*. https://doi.org/10.1155/2021/4376418

CNBC (2021). Consumers lost $56 billion to identity fraud last year—here's what to look out for. https://www.cnbc.com/2021/03/23/consumers-lost-56-billion-dollars-to-identity-fraud-last-year.html

De Aguiar, E. J., Faiçal, B. S., Krishnamachari, B., & Ueyama, J. (2020). A survey of blockchain-based strategies for healthcare. *ACM Computing Surveys (CSUR)*, 53(2), 1–27.

Dhagarra, D., Goswami, M., Sarma, P. R. S., & Choudhury, A. (2019). Big data and blockchain supported conceptual model for enhanced healthcare coverage: The Indian context. *Business Process Management Journal*, 25(7), 1612–1632.

Dimitrov, D. V. (2019). Blockchain applications for healthcare data management. *Healthcare Informatics Research*, 25(1), 51–56.

distilgovhealth.com. (2020). *AI, Machine Learning, and Blockchain are Key for Healthcare Innovation*. [online] Available at: https://distilgovhealth.com/2020/05/27/ai-machine-learning-and-blockchain-are-key-for-healthcare-innovation/ [Accessed 26 Sep. 2024]

Dobson, D. (2018). Convergence of big data, artificial intelligence, and blockchain for competitive advantage. *Competitive Intelligence*, 21(1), 8–15.

Egala, B. S., Pradhan, A. K., Badarla, V., & Mohanty, S. P. (2021). iBlock: an intelligent decentralised blockchain-based pandemic detection and assisting system. *Journal of Signal Processing Systems*, 94(6), 1–14.

Elhoseny, M., Haseeb, K., Shah, A. A., Ahmad, I., Jan, Z., & Alghamdi, M. (2021). IoT solution for AI-enabled privacy-preserving with big data transferring: An application for healthcare using blockchain. *Energies*, 14(17), 5364.

Espeo Blockchain (2019). Lessons to learn from the UAE in blockchain healthcare. https://espeoblockchain.com/blog/blockchain-healthcare

Experian (2015). New Experian Data Quality research shows inaccurate data preventing desired customer insight. https://www.experian.com/blogs/news/2015/01/29/data-quality-research-study/

Farahani, B., Firouzi, F., & Luecking, M. (2021). The convergence of IoT and distributed ledger technologies (DLT): Opportunities, challenges, and solutions. *Journal of Network and Computer Applications*, 177, 102936.

Fatoum, H., Hanna, S., Halamka, J. D., Sicker, D. C., Spangenberg, P., & Hashmi, S. K. (2021). Blockchain integration with digital technology and the future of health care ecosystems: systematic review. *Journal of Medical Internet Research*, 23(11), e19846.

Fintech News (2020). By 2025, blockchain, IoT, machine learning will converge in healthcare. https://www.fintechnews.org/by-2025-blockchain-iot-machine-learning-will-converge-in-healthcare/

Firouzi, F., Farahani, B., Daneshmand, M., Grise, K., Song, J., Saracco, R., Wang, L.L., Lo, K., Angelov, P., Soares, E., & Luo, A. (2021). Harnessing the power of smart and connected health to tackle COVID-19: IoT, AI, robotics, and blockchain for a better world. *IEEE Internet of Things Journal*, 8(16), 12826–12846.

Forbes (2018b). Artificial Intelligence and Blockchain: 3 Major Benefits of Combining These Two Mega-Trends. https://www.forbes.com/sites/bernardmarr/2018/03/02/artificial-intelligence-and-blockchain-3-major-benefits-of-combining-these-two-mega-trends/?sh=686720d74b44

Harmon, S. A., Sanford, T. H., Xu, S., Turkbey, E. B., Roth, H., Xu, Z., ... & Turkbey, B. (2020). Artificial intelligence for the detection of COVID-19 pneumonia on chest CT using multinational datasets. *Nature communications*, 11(1), 1–7.

Hussien, H. M., Yasin, S. M., Udzir, N. I., Ninggal, M. I. H., & Salman, S. (2021). Blockchain technology in the healthcare industry: Trends and opportunities. *Journal of Industrial Information Integration*, 22, 100217.

IBM (2022). Blockchain and artificial intelligence (AI). https://www.ibm.com/topics/blockchain-ai

Jabarulla, M. Y., & Lee, H. N. (2021, August). A blockchain and artificial intelligence-based, patient-centric healthcare system for combating the COVID-19 pandemic: opportunities and applications. *Healthcare*, 9, 1–22.

Jagatheesaperumal, S. K., Mishra, P., Moustafa, N., & Chauhan, R. (2022). A holistic survey on the use of emerging technologies to provision secure healthcare solutions. *Computers & Electrical Engineering*, 99, 107691.

Katuwal, G. J., Pandey, S., Hennessey, M., & Lamichhane, B. (2018). Applications of blockchain in healthcare: current landscape & challenges. *arXiv* preprint arXiv:1812.02776. https://arxiv.org/abs/1812.02776

Khatoon, A., 2020. A blockchain-based smart contract system for healthcare management. *Electronics*, 9(1), 94.

Kuo, T. T., Kim, H. E., & Ohno-Machado, L. (2017). Blockchain distributed ledger technologies for biomedical and health care applications. *Journal of the American Medical Informatics Association*, 24(6), 1211–1220.

Mackey, T.K., Kuo, T.T., Gummadi, B., Clauson, K.A., Church, G., Grishin, D., Obbad, K., Barkovich, R. and Palombini, M., 2019. 'Fit-for-purpose?'–challenges and opportunities for applications of blockchain technology in the future of healthcare. *BMC Medicine*, 17, 1–17.

Maher, M., & Khan, I. (2022). From sharing to selling: challenges and opportunities of establishing a digital health data marketplace using blockchain technologies. *Blockchain in Healthcare Today*, 5, 1–9.

Mamoshina, P., Ojomoko, L., Yanovich, Y., Ostrovski, A., Botezatu, A., Prikhodko, P., ... & Zhavoronkov, A. (2018). Converging blockchain and next-generation artificial intelligence technologies to decentralize and accelerate biomedical research and healthcare. *Oncotarget*, 9(5), 5665–5690.

McGhin, T., Choo, K. K. R., Liu, C. Z., & He, D. (2019). Blockchain in healthcare applications: Research challenges and opportunities. *Journal of Network and Computer Applications*, 135, 62–75.

Mehbodniya, A., Neware, R., Vyas, S., Kumar, M. R., Ngulube, P., & Ray, S. (2021). Blockchain and IPFS integrated framework in bilevel fog-cloud network for security and privacy of IoMT devices. *Computational and Mathematical Methods in Medicine*. https://doi.org/10.1155/2021/7727685

Mehta, N., Pandit, A., & Shukla, S. (2019). Transforming healthcare with big data analytics and artificial intelligence: A systematic mapping study. *Journal of biomedical informatics*, 100, 103311.

Mordor Intelligence (2022). Blockchain market in healthcare - growth, trends, COVID-19 impact, and forecasts (2022–2027). https://www.mordorintelligence.com/industry-reports/blockchain-market-in-healthcare.

Muñoz-La Rivera, F., Mora-Serrano, J., Valero, I. and Oñate, E., 2021. Methodological-technological framework for construction 4.0. *Archives of Computational Methods in Engineering*, 28, 689–711.

Nugent, T., Upton, D., & Cimpoesu, M. (2016). Improving data transparency in clinical trials using blockchain smart contracts. *F1000Research*, 5, 1–7

Onik, M. M. H., Aich, S., Yang, J., Kim, C. S., & Kim, H. C. (2019). Blockchain in healthcare: Challenges and solutions. In Dey, N., Das, H., Naik, B., & Behera, H. S. (Eds.). *Big Data Analytics for Intelligent Healthcare Management* (pp. 197–226). Academic Press.

Pablo, R. G. J., Roberto, D. P., Victor, S. U., Isabel, G. R., Paul, C., & Elizabeth, O. R. (2021). Big data in the healthcare system: a synergy with artificial intelligence and blockchain technology. *Journal of Integrative Bioinformatics*, 1–16. https://doi.org/10.1515/jib-2020-0035.

Rabah, K. (2018). Convergence of AI, IoT, big data and blockchain: A review. *The Lake Institute Journal*, 1(1), 1–18.

Redman, T. C. (2016). Bad data costs the US $3 trillion per year. *Harvard Business Review*, 22, 11–18.

Sandner, P., Gross, J., & Richter, R. (2020). Convergence of blockchain, IoT, and AI. *Frontiers in Blockchain*, 3, 1–5. https://doi.org/10.3389/fbloc.2020.522600

Sharma, P., Borah, M. D., & Namasudra, S. (2021). Improving security of medical big data by using blockchain technology. *Computers & Electrical Engineering*, 96, 107529.

Shae, Z. & Tsai, J.J., 2017, June. On the design of a blockchain platform for clinical trial and precision medicine. In *2017 IEEE 37th international conference on distributed computing systems (ICDCS)*, Atlanta, USA. (pp. 1972–1980). IEEE.

Siyal, A. A., Junejo, A. Z., Zawish, M., Ahmed, K., Khalil, A., & Soursou, G. (2019). Applications of blockchain technology in medicine and healthcare: Challenges and future perspectives. *Cryptography*, 3, 1–16.

Supriya, M., & Chattu, V. K. (2021). A review of artificial intelligence, big data, and blockchain technology applications in medicine and global health. *Big Data and Cognitive Computing*, 5, 1–20.

Tagde, P., Tagde, S., Bhattacharya, T., Tagde, P., Chopra, H., Akter, R., Kaushik, D., & Rahman, M. (2021). Blockchain and artificial intelligence technology in e-Health. *Environmental Science and Pollution Research*, 28(38), 52810–52831.

Tapscott, D. and Tapscott, A., 2020. What blockchain could mean for your health data. *Harvard Business Review*, 12.

WHO (2022). *Patient safety - Data and statistics*. World Health Organization Regional Office for Europe. https://www.who.int/news-room/fact-sheets/detail/patient-safety

Xie, Y., Lu, L., Gao, F., He, S. J., Zhao, H. J., Fang, Y., Yang, J.M., An, Y., Ye, Z. W., & Dong, Z. (2021). Integration of artificial intelligence, blockchain, and wearable technology for chronic disease management: A new paradigm in smart healthcare. *Current Medical Science*, 41(6), 1123–1133.

Yoon, H. J. (2019). Blockchain technology and healthcare. *Healthcare Informatics Research*, 25(2), 59–60.

Chapter 7

An innovative design and simulation of a blockchain-based smart contract framework for enhancing gold traceability

Deo Shao, Martha Shaka, Fredrick Ishengoma, George Bennett, Fredrick Betuel Sawe, and Jay Daniel

7.1 INTRODUCTION

Tracing gold is difficult once mined, it passes through several stakeholders (from miners to consumers), often erasing signatures of origin, resulting in an illegal gold chain. There is serious evidence of illegality in the gold trade. Between 2015 and 2020, for example, Brazil traded an alarming 229 tons that showed "high evidence of illegality". This is almost 50% of the national production. Gold trading is a high-risk industry where consumers increasingly want to know where their gold comes from because of several cases of fraud, duplication, and illegal practices (Neumann et al. 2019; Pochon et. al, 2021; Instituto Escolhas, 2022; Espinosa & Lyons, 2023). While tracking gold is difficult, there are an increasing number of practical solutions that offer a choice between "fully traceable" systems and mass balance systems (where certified gold loses its traceability and transparency within the supply chain but still provides the same benefits to the mining sites themselves) (Finlay, 2020). Recently, blockchain technology is thought to be used to track gold bars in the gold supply chain to ensure authenticity and bolster transparency within the market for the precious metal. By capturing transactions in "blocks" that can be viewed but not modified, blockchain-based ledgers make it easier for buyers to track gold and guarantee its provenance and harder for malicious actors to falsify receipts. The transaction history of each gold bar can be tracked and verified, providing a transparent and unalterable record of the supply chain.

A number of literatures have suggested that blockchain technology can have use cases for such things as avoiding blood gold from conflict zones and illegal organisations. Blockchain represents an alternative system that could increase the traceability of the gold industry which would allow stakeholders to know the provenance of not just each piece of gold in the supply chain, but also how and where it was sourced from, how it was extracted from the land or sea, and what manner of labour was used in the jewellery that it was used to make. With this kind of visibility into the where and how of what it's describing, stakeholders can have more assurance that the solution is one where the integrity and transparency of the entire gold supply chain is beyond reproach (Escolhas, 2022; Mozée, 2022; Quiroz-Gutierrez, 2022; Spence, 2022; Cartier et al., 2018; Mann et al., 2018; Calvão & Archer, 2021).

The blockchain-based decentralised ledger system guarantees data integrity and transparency, regardless of central authority (Rahmani, 2022; Saha et al., 2019; Zikratov et al., 2017), thus redefining the gold trade through unchallengeable transactions. Its advantages marrying with the gold industry include improved tracking of gold from mines to market, assuring ethically sourced offerings and quashing the trade in dirty gold, and intensified security with dramatically lower levels of fraud, as the blockchain simply eliminates the ability to change records, making every piece of gold transparent and traceable from mines to market. As such, throughout its value chain, a gold asset remains genuine and pure, underpinned by the fact that its transaction records cannot be tampered with (Khandelwal, 2019; Mann et al., 2018; Shaikh, 2021). Henceforth, blockchain technology has the potential to revolutionise the industry by addressing challenges such as traceability, transparency, and data provenance.

Nevertheless, there are still a number of challenges. These include technical barriers, regulatory issues, as well as the need for the technology to be adopted all across the sector. Blockchain networks require effective and secure ways of operating high-volume transactions within the gold trading ecosystem (Rao et al., 2024). For integration purposes to ensure a smooth process, it should seamlessly work with the already existing systems in the sector. In addition, lack of explicit rules hampers current attempts to integrate blockchain into the gold market (Buterin et al., 2023). The different regulations on blockchain in different countries confuse actors in the market who do not understand what they should follow. Market participants need clear instructions given under legislation that can assure them of its compliance for proper functioning. To address concerns regarding consumer protection, anti-money laundering measures, and financial stability while promoting innovation, regulators must engage industry stakeholders in collaborative efforts to develop effective policy frameworks consistent with global standards. Overcoming these obstacles will require a collective effort. Regulators, technology providers, and practitioners in the gold market all need to play their part.

7.2 LITERATURE REVIEW

With the ever-increasing global competition, the mining industry is forced to develop and execute supply chain strategies that will enhance its ability to add value to its supply chains and meet the ever-increasing dynamic demand of its key stakeholders (Ada et al, 2021). Furthermore, industries such as the gold supply chain require robust supply chain management systems as success is no longer based on organisations but on their supply chain effectiveness (Garza-Reyes et al, 2017). For this reason, it can be agreed that supply chain is a continuous end-to-end process that helps in integrating organisational workflow from scheduling raw materials all through to the end user (Gozali et al, 2024). Based on this, supply chain management should aim to provide effective management of all activities including transferring the right items to the right customer at the right time and at the lowest cost possible to the ultimate satisfaction of the end users (Shoaib et al, 2023).

It can be argued that supply chain management in the mining industry should enhance its ability to develop and implement value-adding strategies that can effectively integrate the management of all key players (Finlay, 2020). In this context, if an organisation can identify and effectively add value to its customers successfully, they will have established a competitive advantage, guiding the purpose of a strategic supply chain plan (Garza-Reyes et al, 2017). Meanwhile, the gold mining industry faces bottlenecks regarding supply chain visibility and traceability, particularly since stakeholders demand information on the origin of their gold (Ada et al., 2021). In response, organisations have taken various steps to improve the visibility of gold supply chains. Neumann et al. (2019) argue that blockchains may offer a way of improving visibility as well as enhancing traceability within all types of gold supply chains, be it from numerous or few sources. For instance, as shown in Figure 7.1, the applications of blockchain

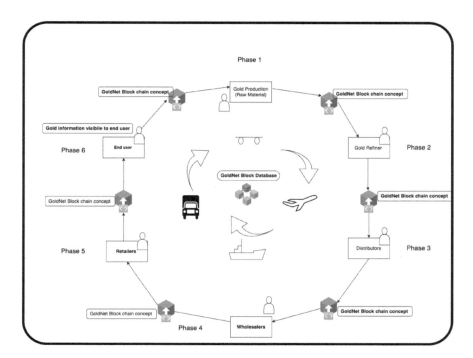

Figure 7.1 Overview of the GoldNet supply chain.

in the gold supply chains can be one of the key factors that can enhance gold chain traceability, promote value creation, and increase customer trust.

Blockchains store transactions in information blocks that are visible but tampering-proof through the use of a decentralised approach. As such, it may be contended that creating blockchain ledgers could boost buyer traceability and solidify safety and trust (Finlay, 2020). For instance, a study on gold mines carried out by Neumann et al. (2019) in some parts of the Democratic Republic of Congo (DRC) revealed that there was an underestimation of gold trade in DRC estimated at $543m to $812m since a considerable portion outputted from mines contained smuggled goods causing huge losses (Finlay, 2020). The study executed four schemes including registering hardware, e-transport bags, electronic ID cards, and online database which uploaded data only to authorised users automatically (Neumann et al., 2019). By so doing, it becomes possible for one to identify any present or future deficiencies within the gold supply chain through resilient traceability measures. Through the implementation of the pilot project, it was possible to trace the movement of about 56% of trade involving exported jewellery from the mining stage; this means that traceability loops could also lead to revenue growth.

In blockchain, the integrity and safety of stored information are assured by one important element – immutability (Ye et al., 2023). Data that goes into the blockchain undergoes cryptographic hashing, once included, changing or tampering of data becomes almost impossible (Pennekamp et al., 2024). This property is possible because of how no single point but many points connected blockchain networks work to prevent such tampering acts. After being confirmed and appended into any block, every transaction gets encrypted and sealed before becoming part of that unalterable record kept by blockchain (Pennekamp et al., 2024).

The tamper-proof feature of blockchain ledgers enhances trust among stakeholders in the gold supply chain such as consumers, regulators, and industry since it is possible to confirm the integrity of data on gold (Schöneich et al., 2023). Through blockchain-based platforms, all players can verify the gold information (Mann et al., 2018; Ahmed, & MacCarthy, 2023). A simple scan of the digital identity would provide details on which mine was the gold from, labour conditions in the mining area, and handling circumstances (Gordon et al., 2023; Yadav et al. 2023). The provided details are crucial in showing how gold moves within the supply chains; it also shows ways through which immediate checks can be carried on whether the gold is genuine or not (Chadly et al., 2023; (Fahim et al., 2023). In this regard, a retailer or wholesaler can make sure that they do not buy unethical or substandard gold by verifying the authenticity of the commodity before buying it. With such advancements, consumers don't need to worry about purchasing a fake or non-compliant gold when using a mobile application for this purpose.

As a transparent solution (Chadly et al., 2023), blockchain holds the potential to promote ethical behaviour and responsibility throughout the entire supply chain since it means that anomalies in an asset's history can easily be spotted, such as gold regarding illegal activities like those mentioned (non-compliant mining, smuggling, and fake certifications). As part of the bigger picture, ethics serve to promote trust among consumers in the gold trade especially, in an age "where products come from, who made them, and what their impact has been" are more important than ever before (Pennekamp et al., 2024).

Various studies have shown that linking the gold market with blockchain technology could improve several aspects of it; transparency, security, efficiency, and trusted gold transactions (Ahram et al., 2017; Ali et al., 2020; Kaur & Gupta, 2021; Lewis et al., 2017; Mermer et al., 2018; Padeli et al., 2022; Silva et al., 2020; Thakker et al., 2021; Zhang, 2020). The amalgamation of blockchain security measures and its decentralised nature makes it highly applicable within the context of the gold industry, where it is proposed, for example, verifying that mining origin and supply chains serve their effective purposes (Gordon et al., 2023; Thakker et al., 2021). This encourages ethical sourcing in the mineral and metal marketplaces (Kshetri, 2022). It has been suggested it could equally lead to continuous surveillance over all stages of the mining process in which commodities are mined (Mann et al., 2018).

7.3 METHODOLOGY

This study applied a Design Science Research (DSR) framework due to the need to develop a solution that addresses a real-world problem (Hevner & Chatterjee, 2010). DSR research is unique because they "do not only make things, but they also answer questions about things and their surroundings" (Kuechler & Vaishnavi, 2008). The DSR approach (where knowledge is produced through the creation of artefacts) is commonly used for such research (March & Smith, 1995). DSR projects typically consist of four phases: "problem construction", "requirements definition", "design and development", and quadruple "demonstration", "evaluation", and "communication" (Johnson & Onwuegbuzie, 2004). A critical literature review was carried out to derive research problem elaboration. The review provided background information as well as insights on the possibility of implementing appropriate digital technology solutions. After that, a proof-of-concept artefact called "GoldNet" was designed based on a standard prototyping software engineering approach. Specifically, the evolutionary prototyping model was used for this purpose. Finally, the ex-ante evaluation approach (Venable et al., 2012) was used to establish the task fit of the developed artefact. Figure 7.2 shows the research process followed in the present study.

134 Blockchain Technology

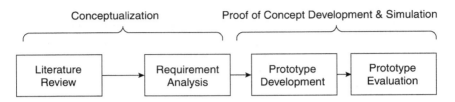

Figure 7.2 Overview of the research process guided this study. (Authors' own construct.)

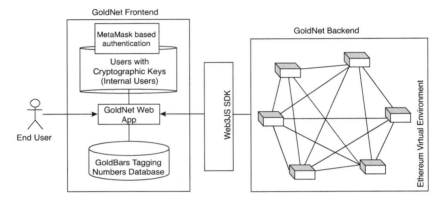

Figure 7.3 GoldNet high-level abstraction. (Authors' own construct.)

7.4 DESIGN OF GOLDNET PROOF-OF-CONCEPT PROTOTYPE

7.4.1 System framework

The GoldNet Prototype is a proof of concept designed from scratch to incorporate blockchain technology within the gold industry. The GoldNet Prototype is implemented using a combination of modern web development frameworks, blockchain development tools, and smart contract programming. In the end, a secure, efficient, and easy-to-use platform is built. The GoldNet Prototype high-level abstraction is depicted in Figure 7.3. The subsequent subsections explain the technology stack of the proposed GoldNet.

7.4.1.1 Frontend development

The dynamic and interactive features of the GoldNet prototype were built using the React[1] web framework. Furthermore, the Vite[2] library is bundled with the React application. This library was chosen because it provides a faster and leaner development experience compared to traditional tools such as Webpack. TypeScript programming language was chosen for the frontend development. TypeScript extends JavaScript by adding types, increasing code quality and maintainability.

Figure 7.4 GoldNet simulation environment. (Authors' own construct.)

Internal users of GoldNet are authenticated using the MetaMask[3] software cryptocurrency wallet. This software allows secure interaction with the Ethereum blockchain, on which they manage their accounts and sign transactions securely (Suratkar et al., 2020). This software is necessary when integrating blockchain features directly into the frontend, as this allows for easily easing user interaction with smart contracts and the blockchain.

7.4.1.2 Backend development

Ganache[4] software framework was used as a local development blockchain environment for testing and development purposes (*see Figure 7.4 – sample screenshot of the blockchain simulation environment setup*). It allows the creation and management of Ethereum blockchain for deploying contracts and developing Decentralised Applications (DApp) (Lee, 2019a). It provides a controlled and safe environment for developers to experiment and debug before deploying to the main network (Singh & Chakraverty, 2022).

The connection between the frontend and backend was established using the Web3JS[5] Software Development Kit (SDK). Web3JS is a library collection that allows interaction with a local or remote Ethereum node over HTTP, IPC, or WebSocket (Lee, 2019b). It enabled the frontend to communicate with the Ganache blockchain, send and receive transactions, interact with smart contracts, and listen to events on the blockchain.

Smart contracts for GoldNet are written in Solidity, the programming language. These contracts are deployed and executed within the Ganache framework (Mohanty, 2018; 2019), facilitating various operations such as gold asset tokenisation, tracking, and verification processes defined in the algorithms.

7.4.2 GoldNet workflow design

Internal actors of the GoldNet system – including Mining Company Officers, Tax Authority Officers, Shipment Officers, and Sales Officers – possess an Ethereum Address (EA). These actors participate actively by invoking functions within the deployed smart contract on the blockchain network. In this way, all transactions are logged securely, transparently, and immutably on the ledger of the blockchain network. For instance, a Mining Company Officer initiates a sequence of function invocations within the smart contract via the Backend Server to tokenise the gold bars on the blockchain.

Upon successful registration and tokenisation, the blockchain network reverts the action to the Backend Server, conveying this success to the Frontend Application, thus informing the Mining Company Officer that the registration is complete. Similarly, the Tax Authority Officer participates in the system through the registration of tax clearance details into the Frontend Application, which is processed by the Backend Server to invoke the relevant smart contract function on the blockchain, thus logging the tax clearance details immutably, in this way, outlining a transparent and secure mechanism of gold transaction taxation.

Subsequently, Shipment Officers and Sales Officers interact with the system by submitting shipment clearance and sales forms, respectively, that are subsequently validated and processed through the smart contract, thus ensuring precision, security, and compliance for regulatory reporting. Finally, the End User or any public actor can authenticate and trace the history of a gold bar through its physical tag via the system. This robust interaction model – enabled by smart contracts – not only enhances the security and transparency of gold trading but also rigorously streamlines the entire gold lifecycle, from extraction and tokenisation to sales and verification. Figure 7.5 further depicts the workflow of events in the GoldNet prototype.

7.4.3 GoldNet smart contract design

A smart contract is an immutable program stored on a blockchain (Vacca et al., 2021). It automates the execution of transactions based on predetermined conditions being met, and they are widely used to execute agreements in a decentralised manner without middlemen. Smart contract greatly reduces the degree of manual participation and ensure the decentralisation of blockchain and the tamper-proof of data. The proposed GoldNet traceability system based on blockchain designed in this paper records traceability data of gold from mine, processing, shipping, and sales.

The proposed GoldNet comprises six business processes: user registration management, gold tokenisation management, tax clearance management, gold shipment management, gold sales management, and gold history and provenance traceability. Figure 7.6 shows a schematic diagram of the GoldNet Processes discussed in the subsequent subsections.

7.4.3.1 Gold extraction and tokenisation

Mining companies initiate the process by registering gold bars on the blockchain, providing detailed information to ensure traceability and compliance. Algorithm 1 illustrates the pseudocode of the gold tokenisation process. Figure 7.7 shows the gold registration and tokenisation management page.

Blockchain-based smart contract simulation framework 137

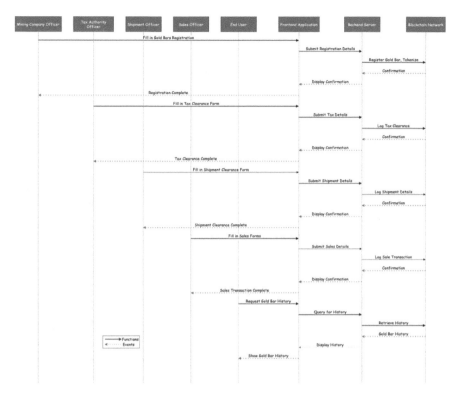

Figure 7.5 Sequence diagram showing interactions among actors. (Authors' own construct.)

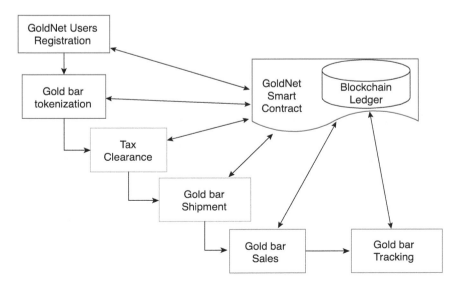

Figure 7.6 Schematic diagram of the GoldNet processes. (Authors' construct.)

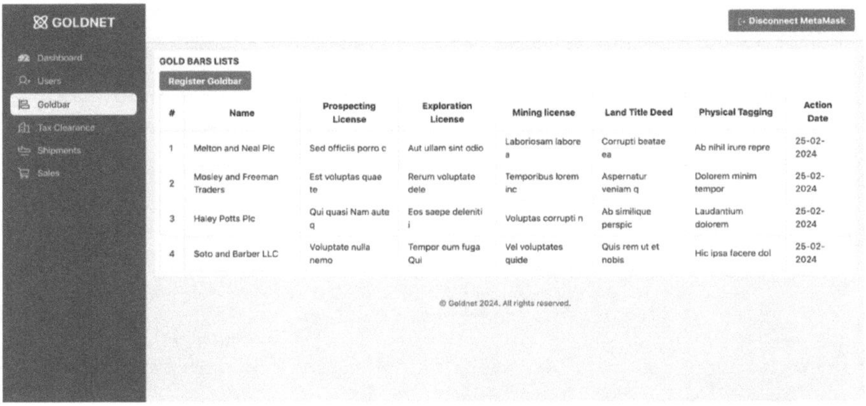

138 Blockchain Technology

Figure 7.7 Gold tokenisation. (Authors' own construct.)

Algorithm 1: Register and Tokenisation of Gold Bar

1. **Input:** goldBarRegistrationData: {CompanyName,DistinctiveCharacteristics, ChemicalUsed, Dosage, ProspectingLicense, ExplorationLicense, MiningLicense, LandTitleDeed, PhysicalTagging, ActionDate}
2. **Output:** goldBarTokenID
3. **Begin**
4. **Validate** goldBarRegistrationData
5. **Generate** goldBarTokenID linked with goldBarRegistrationData
6. **Add** goldBarToken to GoldNet with ActionDate
7. **Return** goldBarTokenID
8. **End**

7.4.3.2 Taxation clearance

The tax authority completes the tax clearance form for the gold, ensuring that all tax obligations are met before any further actions. Algorithm 2 illustrates the pseudocode of the gold tax clearance process. Figure 7.8 shows a sample screenshot of the gold tax clearance management page.

Figure 7.8 Gold tax clearance. (Authors' own construct.)

Algorithm 2: Tax Clearance

1. **Input:** taxClearanceData:{TaxClearanceID, Description, ActionDate}
2. **Output:** Update GoldNet with taxClearanceData
3. **Begin**
4. **Record** taxClearanceData on GoldNet
5. **End**

7.4.3.3 Gold shipment

For gold to be shipped, the mining company or dealer must fill in the shipment clearance form, including necessary permits. Algorithm 3 illustrates the pseudocode of the gold shipment process. Figure 7.9 shows a sample screenshot of the gold shipment management page.

Algorithm 3: Gold Shipment

1. **Input:** shipmentClearanceData: {CentralBankAuthorization, SecurityClearanceID, ElectronicInvoicingID, ShipmentPermit, ActionDate}
2. **Output:** Update GoldNet with shipmentClearanceData
3. **Begin**
4. **Validate** shipmentClearanceData
5. **Record** shipmentClearanceData on GoldNet
6. **End**

Figure 7.9 Gold shipment management. (Authors' own construct.)

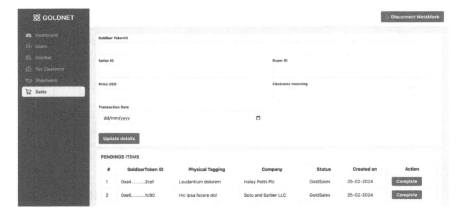

Figure 7.10 Gold sales management. (Authors' own construct.)

7.4.3.4 Gold sales

The sale of gold is recorded on the blockchain, detailing the transaction between the seller and buyer, including the price and date. Algorithm 4 illustrates the pseudocode of the gold sales process. Figure 7.10 shows a sample screenshot of the gold sales management page.

Algorithm 4: Gold Sales

1. **Input:** salesData: {SellerID, BuyerID, PriceUSD, TransactionDate, ElectronicInvoicingID}
2. **Output:** Update GoldNet with salesData
3. **Begin**
4. **Record** salesData on GoldNet
5. **End**

7.4.3.5 Third-Party Tracking

Public actors can verify and trace the history of a gold bar using its physical tag, ensuring transparency and accountability. Algorithm 5 illustrates the pseudocode of the gold tracking process. Figure 7.11 depicts the sample screenshot of the gold bar history and provenance tracking.

Algorithm 5: Track Gold History

1. **Input:** physicalTaggingID
2. **Output:** goldBarHistory
3. **Begin**
4. **Retrieve** goldBarHistory using physicalTaggingID from GoldNet
5. **Return** goldBarHistory
6. **End**

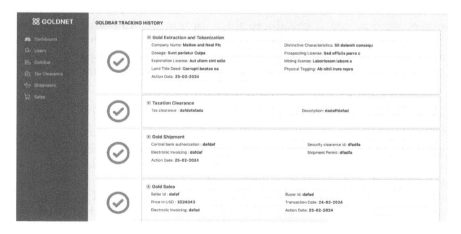

Figure 7.11 Gold history and provenance tracking. (Authors' own construct.)

7.5 DISCUSSION

Blockchain technology provides an opportunity to address the gold industry's critical transparency, efficiency, and ethical compliance challenges. As a result, it can revolutionise traditional practices, offering enhanced traceability, secured transactions, and fostering an equitable trading environment. This study designed and implemented a GoldNet showcasing the traceability of gold from its extraction to retail. The proposed GoldNet framework can potentially improve the gold trading system with far-reaching practical and policy implications, as discussed below.

7.5.1 Practical implications

Our work developing and testing a blockchain system for the gold supply chain offers valuable insights into integrating blockchain technology with ongoing gold traceability efforts. For instance, the DRC has undertaken initiatives to enhance gold supply chain oversight through an Electronic Traceability System. In this system (Filipova et al., 2018), a range of technological tools are used, including smartphones and tablets for registration, gold transport bags for electronic reading, electronic ID cards for access control, and a centralised data repository. Although the pilot in the DRC showed that the system has the potential to increase transparency and to allow the legal trading of gold to be managed effectively, it also showed many vulnerabilities, such as that the record of transactions can be altered. Our simulation results make a case in favour of the option of seriously considering blockchain as a robust solution in future pilots that aim to increase the traceability of gold.

In sum, in the proposed framework, the capability to validate the provenance and legitimacy of gold promotes a trustworthy relationship among buyers and sellers, which is in line with the development of an ethically sourced market that is friendly to the environment. Additionally, the tax clearance proposed by the framework eases the process, which can then promote the collection of taxes that is both efficient and transparent, therefore supporting the process of nations and the hold of accountability.

7.5.2 Policy implications

Advancing blockchain in gold trading requires consideration of political, technical, and financial dimensions, which are essential. Policy-driven initiatives should promote innovation, infrastructure growth, and a nurturing technological, social, and economic development ecosystem. For instance, in the SADC Industrialisation Strategy and Roadmap for 2015– 2063, one of the critical strategies proposed is developing viable regional value chains (SADC report, 2015). Our simulation involved various stakeholders, such as mining, tax, trade, and financial experts. Through their interaction with the systems, the regulatory organisation can observe and devise novel strategies which can later be developed into policies, e.g., regional competitiveness on a global level, which aligns with one of the Mineral Beneficiation Strategies in the SADC's industrial strategy. Also, blockchain standardises trading practices, facilitating integration with international markets.

7.5.3 Future directions

The research here paves the way for further studies. A potential future research focus could be the empirical evaluation of how blockchain can mitigate illegal gold trading while bolstering ethical sourcing practices. The empirical assessment can give insight into the effectiveness of adopting blockchain and areas for further enhancement, for example, analysing the effect of blockchains on the market dynamics and customer behaviour due to the enhanced trust and verification of gold sourcing. There is also an opportunity for researchers to explore the synergy between blockchain and emerging technologies such as the Internet of Things (IoT) and artificial intelligence (AI) to advance traceability in other industries (Charles et al. 2023). Furthermore, the development of supporting regulatory frameworks and international collaboration that facilitates blockchain adoption in gold trading is needed.

7.6 CONCLUSION

This chapter introduces a tactical roadmap to the deployment of blockchain technology into mining supply chain of gold trade. The objective is to offer insights about the reengineering of supply chain operations in gold trade with the introduction of blockchain. The GoldNet simulation demonstrates that the incorporation of blockchain technology into the gold trade supply chain marks the advent of a new era in the trading of commodities. Further, the insights from this study contribute to improved supply chain operations for gold and provide actionable insights for blockchain adoption. By implementing blockchain across the gold supply chain, stakeholders can achieve levels of transparency and trust that are unprecedented, and by doing so they will set a new standard that the industry can't ignore.

This chapter also highlights the importance of bringing together multiple players in order to leverage all that blockchain technology has to offer. Improve how blockchain is used, leading to a more efficient, transparent gold supply chain that eliminates the ongoing risks and uncertainties that we're so used to. As technology advancement keeps advancing, the combined efforts of government, industry, academia, and community groups are necessary in order to assure efficient, transparent, ethically, environmentally friendly gold supply chain. Therefore, by overcoming the initial difficulties of working with this new technology and using it as a foundation to craft and enact policy while performing necessary research, players can take maximum advantage of blockchain, not only resulting in sustainable practices but also bringing teamwork and innovation along for the ride.

NOTES

1. React: https://react.dev/.
2. Vite: https://vitejs.dev/.
3. MetaMask: https://metamask.io/.
4. Ganache: https://trufflesuite.com/ganache/.
5. Web3js: https://web3js.org/.

REFERENCES

Ada, N., Ethirajan, M., Kumar, A., K.E.K, V., Nadeem, S. P., Kazancoglu, Y. and Kandasamy, J. (2021) 'Blockchain technology for enhancing traceability and efficiency in automobile supply chain—a case study', *Sustainability*, 13(24), p. 13667. doi: 10.3390/su132413667.

Ahmed, W. A., & MacCarthy, B. L. (2023). Blockchain-enabled supply chain traceability–How wide? How deep? *International Journal of Production Economics*, 263, 108963.

Ahram, T., Sargolzaei, A., Sargolzaei, S., Daniels, J., & Amaba, B. (2017, June). Blockchain technology innovations. *2017 IEEE Technology & Engineering Management Conference (TEMSCON)*, Santa Clara, CA, USA, 2017, pp. 137–141, doi: 10.1109/TEMSCON.2017.7998367.

Ali, O., Ally, M. A., Clutterbuck, P., & Dwivedi, Y. K. (2020). The state of play of blockchain technology in the financial services sector: A systematic literature review. *Int. J. Inf. Manag.*, 54, 102199.

Buterin, V., Illum, J., Nadler, M., Schär, F., & Soleimani, A. (2023). Blockchain privacy and regulatory compliance: Towards a practical equilibrium. *Blockchain: Research and Applications*, 100176.

Calvão, F., & Archer, M. (2021). Digital extraction: Blockchain traceability in mineral supply chains. *Political Geography*, 87, 102381.

Cartier, L. E., Ali, S. H., & Krzemnicki, M. S. (2018). Blockchain, chain of custody and trace elements: An overview of tracking and traceability opportunities in the gem industry. *Journal of Gemmology*, 36(3).

Chadly, A., Hasan, H. R., Moawad, K., Salah, K., Omar, M., & Mayyas, A. (2023). A blockchain-based solution for the traceability of rare earth metals used in thin-film photovoltaics. *Journal of Cleaner Production*, 428, 139399.

Charles, V., Emrouznejad, A., & Gherman, T. (2023). A critical analysis of the integration of blockchain and artificial intelligence for supply chain. *Annals of Operations Research*, 327, 7–47. https://doi.org/10.1007/s10479-023-05169-w

Espinosa, C., & Lyons, C., 2023. Can blockchain clean up the gold industry in Brazil and across the globe? Mongabay. https://news.mongabay.com/2023/11/can-blockchain-clean-up-the-gold-industry-in-brazil-and-across-the-globe (Accessed 16 February 2024).

Fahim, S., Rahman, S. K., & Mahmood, S. (2023). Blockchain: A Comparative study of consensus algorithms PoW, PoS, PoA, PoV. *International Journal of Mathematics and Computer Science*, 3, 46–57.

Filipova, Nadezhda. (2018). *Blockchain – An Opportunity For Developing New Business Models,* Business Management, D. A. Tsenov Academy of Economics, Svishtov, Bulgaria, issue 2 Year 20, pp. 75–92.

Finlay, D., 2020. The 'burden' of traceability in gold supply chains. *Journal of Fair Trade*, 2(1), 22–26. https://doi.org/10.13169/jfairtrade.2.1.0022

Garza-Reyes, J. A., Kumar, V., Martinez-Covarrubias, J. L. and Lim, M. K. (2017) Managing innovation and operations in the 21st century. 1. https://doi.org/10.4324/9781315116860.

Gordon, S. R., Gilleran, R., Shankaranarayanan, G., Stoddard, D. B., & Johnson, B. A. (2023). Blockchain adoption in the supply chain – a game theoretic perspective: The case of the diamond industry. *Journal of Information Technology Case and Application Research*, 25(3), 241–276.

Gozali, L., Kristina, H. J., Yosua, A., Zagloel, T. Y. M., Masrom, M., Susanto, S., Tanujaya, H., Irawan, A. P., Gunadi, A., Kumar, V., Garza-Reyes, J. A., Jap, T. B. and Daywin, F. J. (2024) The improvement of block chain technology simulation in Supply Chain Management (case study: Pesticide company), *Nature News. Nature Publishing Group*. Available at: https://www.nature.com/articles/s41598-024-53694-w (Accessed: 3 March 2024).

Hevner, A., & Chatterjee, S. (2010). *Design research in information systems: Theory and practice* (Vol. 22). Springer Science & Business Media.

Instituto Escolhas. (2022). Blockchain, Traceability, and Monitoring for Brazilian Gold. https://escolhas.org/wp-content/uploads/Blockchain-traceability-and-monitoring-for-Brazilian-gold.pdf (Accessed 16 February 2024)

Johnson, R. B., & Onwuegbuzie, A. J. (2004). Mixed methods research: A research paradigm whose time has come. *Educational Researcher*, 33(7), 14–26.

Kaur, M., & Gupta, S. (2021). Blockchain Technology for Convergence: An Overview, Applications, and Challenges. In S. Pani, S. Lau, & X. Liu (Eds.), *Blockchain and AI Technology in the Industrial Internet of Things* (pp. 1–17). IGI Global. https://doi.org/10.4018/978-1-7998-6694-7.ch001.

Khandelwal, S. (2019). Blockchain technology: Heart of digital financial infrastructure for managing trust and governance system (January 5, 2019). Proceedings of 10th International Conference on Digital Strategies for Organizational Success, Available at SSRN: https://ssrn.com/abstract=3308578 or http://dx.doi.org/10.2139/ssrn.3308578.

Kshetri, N. (2022). Blockchain systems and ethical sourcing in the mineral and metal industry: A multiple case study. *The International Journal of Logistics Management*, 33(1), 1–27.

Kuechler, B., & Vaishnavi, V. (2008). On theory development in design science research: Anatomy of a research project. *European Journal of Information Systems*, 17(5), 489–504. https://doi.org/10.1057/ejis.2008.40

Lee, W.-M. (2019a). Testing smart contracts using ganache. In *Beginning Ethereum Smart Contracts Programming: With Examples in Python, Solidity, and JavaScript* (pp. 147–167). Apress.

Lee, W.-M. (2019b). Using the web3.js APIs. In *Beginning Ethereum Smart Contracts Programming: With Examples in Python, Solidity, and JavaScript* (pp. 169–198). Apress.

Lewis, R., McPartland, J., & Ranjan, R. (2017). Blockchain and financial market innovation. *Economic Perspectives*, 41(7), 1–17.

Mann, S., Potdar, V., Gajavilli, R. S., & Chandan, A. (2018). Blockchain technology for supply chain traceability, transparency and data provenance. *Proceedings of the 2018 International Conference on Blockchain Technology and Application*. https://api.semanticscholar.org/CorpusID:59604259

March, S. T., & Smith, G. F. (1995). Design and natural science research on information technology. *Decision Support Systems*, 15(4), 251–266.

Mermer, G. B., Zeydan, E., & Arslan, S. S. (2018, May). An overview of blockchain technologies: Principles, opportunities and challenges. *2018 26th Signal Processing and Communications Applications Conference (SIU)*, (pp. 1–4). IEEE. Izmir, Turkey.

Mohanty, D. (2018). Deploying smart contracts. In *Ethereum for Architects and Developers* (pp. 105–138). Apress Berkeley, CA. https://doi.org/10.1007/978-1-4842-4075-5.

Mohanty, D. (2019). Supply Chain—Gold Tokenization. In: R3 Corda for Architects and Developers (pp. 193–198), Apress, Berkeley, CA. https://doi.org/10.1007/978-1-4842-4529-3_11

Mozée, C., 2022. The gold industry is testing blockchain technology to track the global market's supply chain of bars. *Business Insider*. https://markets.businessinsider.com/news/commodities/gold-blockchain-bars-crypto-supply-chain-bullion-metals-market-wgc-2022-3 (Accessed 16 February 2024)

Neumann, M., Barume, B., Ducellier, B., Ombeni, A., Näher, U., Schütte, P., von Baggehufwudt, U., Ruppen, D., & Weyns, Y., 2019. Traceability in artisanal gold supply chains in the Democratic Republic of the Congo: Lessons Learned from the Kampene Gold Pilot Project. https://www.bgr.bund.de/DE/Themen/Min_rohstoffe/Downloads/studie_traceability_in_artisanal_gold_DR_Congo_2019.pdf?__blob=publicationFile&v=10 (Accessed 23 February 2024)

Padeli, P., Husain, A., & Julianto, D. P. (2022). Gold-Based financial information system design using blockchain application. *Blockchain Frontier Technology*, 2(1), 9–16.

Pochon, A., Desaulty, A.-M., Bailly, L. and Lach, P. (2021) 'Challenging the traceability of natural gold by combining geochemical methods: French Guiana example', *Applied Geochemistry*, 129, p. 104952. doi: 10.1016/j.apgeochem.2021.104952.

Pennekamp, J., Alder, F., Bader, L., Scopelliti, G., Wehrle, K., & Mühlberg, J. T. (2024). *Securing Sensing in Supply Chains: Opportunities, Building Blocks, and Designs*. IEEE Access.

Quiroz-Gutierrez, M., 2022. The gold industry sees blockchain as a salvation for the bullion black market. *Yahoo*. https://finance.yahoo.com/news/gold-industry-sees-blockchain-salvation-215322912.html (Accessed 25 February 2024).

Rao, I. S., Kiah, M. L., Hameed, M. M., & Memon, Z. A. (2024). Scalability of blockchain: a comprehensive review and future research direction. *Cluster Computing*, 1–24. doi: 10.1007/s10586-023-04257-7

Rahmani, M. K. I. (2022). Blockchain technology: principles and algorithms. In Shahnawz, K., Syed, M. H., Hammad, R. and Bushager, A. F. (Eds.), *Blockchain Technology and Computational Excellence for Society 5.0* (pp. 16–27). IGI Global.

SADC report (2015). *SADC Industrialization Strategy and Roadmap*. SADC https://www.sadc.int/sites/default/files/2022-07/Repriting_Final_Strategy_for_translation_051015.pdf (Accessed 20 February 2024).

Saha, S., Jana, B., & Poray, J. (2019). A Study on Blockchain Technology. CompSciRN: Other Cybersecurity. https://api.semanticscholar.org/CorpusID:219377327

Schöneich, S., Saulich, C., & Müller, M. (2023). *Traceability and Foreign Corporate Accountability in Mineral Supply Chains.* Regulation & Governance.

Shaikh, A. (2021). Overview of Blockchain Technology in current Times. *International Journal of Advanced Research in Science, Communication and Technology,* 331–338. https://ijarsct.co.in/Paper1652.pdf

Shoaib, M., Zhang, S., & Ali, H. (2023). A bibliometric study on blockchain-based supply chain: a theme analysis, adopted methodologies, and future research agenda. *Environmental Science and Pollution Research, 30*(6), 14029–14049.

Silva, T. B. da, Morais, E. S. de, Almeida, L. F. F. de, Rosa Righi, R. da, & Alberti, A. M. (2020). Blockchain and industry 4.0: Overview, convergence, and analysis. In *Blockchain Technology for Industry 4.0: Secure, Decentralized, Distributed and Trusted Industry Environment* (pp. 27–58). Springer, Singapore.

Singh, S., & Chakraverty, S. (2022, December 21). Implementation of proof-of-work using Ganache. *2022 IEEE Conference on Interdisciplinary Approaches in Technology and Management for Social Innovation (IATMSI),* Gwalior, India, pp. 1–4, doi: 10.1109/IATMSI56455.2022.10119271.

Spence, E., 2022. Gold industry embraces blockchain to secure its supply chain. *Bloomberg.* https://www.bnnbloomberg.ca/gold-industry-embraces-blockchain-to-secure-its-supply-chain-1.1744118 (Accessed 25 February 2024)

Suratkar, S., Shirole, M., & Bhirud, S. (2020, September 28). Cryptocurrency wallet: A review. *2020 4th International Conference on Computer, Communication and Signal Processing (ICCCSP),* Chennai, India, 2020, pp. 1–7, doi: 10.1109/ICCCSP49186.2020.9315193.

Thakker, U., Patel, R., Tanwar, S., Kumar, N., & Song, H. (2021). Blockchain for diamond industry: opportunities and challenges. *IEEE Internet of Things Journal, 8*(11), 8747–8773.

Vacca, A., Di Sorbo, A., Visaggio, C. A., & Canfora, G. (2021). A systematic literature review of blockchain and smart contract development: Techniques, tools, and open challenges. *Journal of Systems and Software, 174,* 110891.

Venable, J., Pries-Heje, J., & Baskerville, R. (2012). A comprehensive framework for evaluation in design science research. *Design Science Research in Information Systems. Advances in Theory and Practice: 7th International Conference, DESRIST 2012,* Las Vegas, NV, USA, May 14–15, 2012. Proceedings 7, 423–438.

Yadav, A. K., Singh, K., Amin, A. H., Almutairi, L., Alsenani, T. R., & Ahmadian, A. (2023). A comparative study on consensus mechanism with security threats and future scopes: Blockchain. *Computer Communications, 201,* 102–115.

Ye, T., Luo, M., Yang, Y., Choo, K. K. R., & He, D. (2023). A survey on redactable blockchain: Challenges and opportunities. *IEEE Transactions on Network Science and Engineering, 10*(3), 1669–1683.

Zhang, J. (2020). Deploying blockchain technology in the supply chain. Thomas, C., Fraga-Lamas, P., & M. Fernández-Caramés, T. (Eds.). (2020), In *Computer Security Threats.* IntechOpen, doi: 10.5772/intechopen.86530

Zikratov, I., Kuzmin, A., Akimenko, V., Niculichev, V., & Yalansky, L. (2017, April). Ensuring data integrity using blockchain technology. *2017 20th Conference of Open Innovations Association (FRUCT),* St.Petersburg, Russia, 2017, pp. 534–539, doi: 10.23919/FRUCT.2017.8071359.

Part 4

Managing the blockchain technology

Chapter 8

Managing and planning blockchain implementation framework

A conceptual model

Jay Daniel and Jehan Zaib

8.1 INTRODUCTION

Blockchain technology is widely recognized as the fifth disruptive innovation in computer science. It provides a reliable and transparent way of recording and transferring information. The system maintains a public ledger of all digital transactions, accessible to all parties involved. Each transaction is recorded in an immutable, decentralized manner and certified by all contributors involved in the transaction (Saraji, 2023).

The concept of distributed ledger technology has a lengthy history, dating back to the publication of "New Directions in Cryptography" by Diffie and Hellman in 1976 (Diffie & Hellman, 2022). Stuart Haber and Scott Stornetta later proposed utilizing this concept to timestamp data in their 1991 paper, "How to Time-Stamp a Digital Document" (Haber & Stornetta, 1991). However, it wasn't until Satoshi Nakamoto's 2008 paper, "Bitcoin: A Peer-to-Peer Electronic Cash System," that the system's practical application was envisioned (Nakamoto, 2008). In this paper, Nakamoto (2008) introduced an electronic payment system based on cryptographic proof instead of trust, enabling direct transactions between two parties without the need for a third party. Sellers would be protected from fraud by transactions that are computationally impractical to reverse, while buyers could be safeguarded by routine escrow mechanisms.

Ever since Bitcoin became popular, blockchain technology has been shaking things up in various industries. Many of them are interested in using it to improve their transactional transparency, efficiency, and affordability (Anandan & Deepak, 2020). Researchers have examined the possibilities of implementing this technology in different fields, considering its unique features, such as data security, confidentiality, and integrity. They are motivated to explore the unique attributes of blockchain, such as data security, secrecy, and integrity, to evaluate the impact it can have on the operations of various industries (Singh et al., 2021).

Organizations engage in various transactions with clients and other entities in today's business world. These transactions may involve financial transactions or the sharing of sensitive information, depending on the level of interaction. It is crucial for both parties to trust each other during these transactions without any audit or verification mechanism in place. Blockchain technology can help establish this trust between parties by enabling them to conduct transactions on a secure, transparent, and immutable ledger distributed across multiple servers. This added layer of security and transparency can help organizations build the trust needed for successful business transactions (Swan, 2015).

The advent of blockchain technology has revolutionized the way transaction parties conduct business. This advanced technology offers a highly secure method for storing and transmitting transactional data, eliminating the risk of fraud or manipulation. However, the application of this technology can vary depending on the industry (Jena & Dash, 2021). Therefore, it is crucial for organizations to evaluate how blockchain technology affects their transactional processes and determine the most effective approach to incorporating it into their workflows.

This chapter proposes a comprehensive framework to assist organizations with effectively managing, implementing, and evaluating the performance of blockchain technology in practice. This framework considers the unique characteristics and requirements of various industries and provides a structured approach to evaluate the effectiveness of blockchain technology in practical applications. By utilizing this framework, organizations can optimize the potential of this innovative technology to enhance their operational process transparency, security, and efficiency.

Developing the proposed framework was essential to overcome certain limitations that have impeded the practical application of blockchain technology in various industries. However, a limited amount of research has been conducted to gauge the true impact of these limitations on real-world blockchain use cases. Namasudra et al., (2021) has shed light on several disadvantages that could prevent organizations from adopting this technology in their transactions, including issues with infrastructure support, blockchain functionality, and lack of available standards (Namasudra et al., 2021).

One notable drawback of the system is its tendency toward wastefulness, given that each node essentially duplicates the same process, leading to decreased efficiency and increased time consumption. Additionally, the required cost and network speed to avoid backlogs can be substantial, as more powerful nodes are prioritized over weaker ones, resulting in potential backlogs for the latter.

The implementation of blockchain technology poses particular challenges due to its basic functionality. The size of both the blocks and chains can cause a significant increase in the database's size. Each block must be added to the blockchains of every node, potentially leading to sluggishness and complexity compared to a centralized system.

As blockchain technology is still in its nascent stages, there currently needs to be set standards in place. This can pose a challenge when integrating the technology into an organization's workflow. Given the hurdles mentioned above, it is imperative to establish a framework that evaluates the impact of these challenges on the practical implementation of blockchain technology.

This chapter aims to provide a detailed solution to the challenges faced by industries when adopting blockchain technology in their transaction processes. To address these challenges, this chapter proposes an all-encompassing general framework that any sector can adopt. The framework can be tailored to the industry's specific needs, ensuring seamless integration of blockchain technology into their existing processes.

The framework also provides a way to measure the performance of blockchain technology on the industry's operational efficiency. This performance evaluation can help the industry identify improvement areas and make necessary adjustments to optimize the use of blockchain technology. This comprehensive framework will offer a reliable and efficient solution for industries looking to adopt blockchain technology and improve their operational efficiency and transparency.

The next section of this chapter will provide an in-depth analysis of the literature related to blockchain technologies. It will delve into the extensive research conducted by scholars to measure the performance of different types of blockchain, including public, private, and hybrid. The section will also discuss the various models proposed in the literature to measure blockchain performance in practice, such as consensus algorithms, throughput, scalability, and latency. Moreover, it will highlight the limitations of existing technologies and identify the research gaps that need to be addressed to enhance the performance of blockchain technologies. Literature review and framework will be followed by discussion and the conclusion of this chapter.

8.2 LITERATURE REVIEW

The disruptive and popular nature of blockchain technology can be attributed to three key advantages: transaction transparency, efficient processing, and cost reduction. These benefits stem from the technology's core functional attributes of immutability and decentralization. However, organizations must assess the impact of blockchain technology on their unique needs by measuring performance in real-life situations. This will allow for necessary operational improvements to enhance transparency and efficiency.

8.2.1 Evolution of blockchain technology

Blockchain technology is increasingly becoming popular in different sectors, including finance, healthcare, and supply chain. When combined with Internet of Things (IoT), it unlocks even more potential applications of the technology (Dabbagh et al., 2019). Over the years, as adoption has grown, blockchain technology has evolved and been customized to meet the unique requirements of various industries since the emergence of Bitcoin. Each generation of the technology offers distinct advantages over its predecessor (Saxena et al., 2021).

8.2.1.1 Blockchain 1.0

The first generation of blockchain technologies initially focused on the finance industry and gave rise to the now-famous cryptocurrency. Before blockchain, internet transactions relied on intermediary organizations, a time-consuming, risky, and expensive process. The emergence of blockchain technology offered a solution to the challenges of trust, time, and cost by utilizing established technologies like P2P networks, hashing, and encryption. Satoshi Nakamoto, the mastermind behind this innovative technology, proposed Bitcoin to solve these challenges. The resulting technology is irreversible and tamper-proof, with additional layers of security provided by the P2P network. Blockchain technology has since revolutionized how transactions are carried out, providing a fast, secure, and cost-effective means of conducting transactions without intermediaries (Hamdi et al., 2023).

The advent of Bitcoin has brought about a groundbreaking technology that has revolutionized the way we share and verify transactional information. The Blockchain mechanism that supports it enables participants in peer-to-peer networks to confirm and add transaction details to existing blocks collectively. These blocks cannot be altered once added, providing the highest level of security and transparency (Lee, 2019). Cryptocurrencies like Bitcoin and Litecoin allow people to send digital funds across an online network without incurring multiple fees. This opens up a world of possibilities for sending money across geographic and monetary boundaries and to those who do not have a bank account. This innovative mechanism has also been adapted for other industries beyond cryptocurrency (Mignon, 2019).

Bitcoin presents a multitude of benefits compared to conventional payment methods. By eliminating intermediaries from the transaction process, Bitcoin effectively minimizes transaction costs and simplifies the experience, providing users with a more cost-efficient solution. Its P2P networking and immutable ledger characteristics also guarantee secure and transparent transactions. Moreover, these features enable Bitcoin to provide anonymity for users, ensuring their privacy. Lastly, the growing adoption of cryptocurrency payments by large corporations and across various industries further cements Bitcoin's position as an increasingly favored payment option (Dumitrescu, 2017).

Although blockchain technology has benefits within the financial sector, the first generation of this technology has some limitations compared to traditional payment methods. The primary issue is the mining mechanism, which uses algo-puzzles known as proof-of-work (PoW) to confirm and complete transactions (Böhme et al., 2015). Over time, these puzzles have become increasingly difficult to solve, resulting in longer wait times and occasional failed attempts (Bhushan et al., 2022). Additionally, the PoW system requires more infrastructural support and time to mine the bitcoins, resulting in a slow throughput of only around seven transactions per second, compared to credit card companies that can handle roughly 2000 transactions per second (Vukolić, 2016). This scalability issue directly affects the widespread adoption of the technology. Böhme et al., (2015) also noted the oversized mining pool of Bitcoin, which creates a vulnerability to attacks and the potential for false transactions.

8.2.1.2 Blockchain 2.0

The widespread adoption of Bitcoin has been hindered by two significant limitations: excessive energy consumption(De Vries, 2018) and limited scalability (Malik et al., 2022). However, these limitations have spurred computer science researchers and programmers to seek improvements. Among these pioneers is Vitalik Buterin, a Russian-Canadian programmer who introduced the concept of the second generation of Blockchain. Today, Ethereum stands as the most widely used Blockchain 2.0 platform (Buterin, 2013).

Upon its introduction, Ethereum garnered the attention of many individuals interested in utilizing cryptocurrencies for financial contracts, including "contracts for difference," where two parties agree to exchange funds based on the value of a specific asset. However, Ethereum took this concept to the next level by introducing contracts that function like miniature computer programs within the blockchain. Each contract features its distinct code executed whenever someone sends a transaction. This code can access valuable information regarding the transaction, such as the sender and the amount involved, and can even initiate its transactions. Essentially, Ethereum contracts are self-sufficient agents within the blockchain that can conduct actions automatically based on the rules outlined within their code (Buterin, 2022).

The capabilities of Blockchain 2.0 offer great potential for scalability and application development in decentralized systems (DApps) (Johnston et al., 2014). Platforms like Ethereum provide a secure layer of automated protection, allowing parties to agree on set terms and ensuring that they are met upon completion of the transaction (Zarir et al., 2021). This technology is known for its cost-efficient operations, as well as its ability to record data immutably, provide transparency throughout the transaction process, and eliminate the need for intermediaries. Furthermore, Ethereum enables direct participation from involved parties, making contract agreements more accessible and streamlined.

Before using smart contract systems for transaction processing and business dealings, it's crucial to be aware of their limitations despite their potential advantages. Delmolino et al., (2016) recounted his experience instructing undergraduate students at the University of Maryland in smart contract programming and emphasized that the intricate coding required in drafting contracts is a potential obstacle for smart contract systems (Delmolino et al., 2016). Even minor coding errors can result in unintended consequences for the parties involved in the contract (Chen et al., 2017).

Similar to its predecessor, the second generation of blockchain technology has faced challenges due to the lack of international regulations (Mitrofanova, 2018) and potential security breaches that could impact smart contracts (Chen et al., 2023). While it is more efficient and faster than the first generation, it still needs to match the transaction processing speed of centralized payment methods like Visa and PayPal. Furthermore, like Blockchain 1.0, Ethereum also uses "Proof-of-Work (PoW)" as the consensus protocol, therefore it does not offer a comprehensive platform with desired scalability, interoperability, sustainability, privacy, and governance functionalities. As a result, Blockchain 2.0 has been an inadequate, insecure, and costly solution, leading to the development of Blockchain 3.0 through notable projects (Singh & Malhotra, 2023).

8.2.1.3 Blockchain 3.0

The initial two iterations of Blockchain technology have encountered a considerable hurdle in scalability. This arises from their use of Proof-of-Work (PoW) as a consensus protocol, necessitating substantial resource allocation. Moreover, PoW is a complex undertaking that often leads to a transaction approval time of approximately 60 minutes, further impeding scalability for both generations (Auer, 2019). As a result, an advanced version of Blockchain 2.0 was proposed to tackle the scalability, sustainability, security, cost, and interoperability concerns related to the first two generations of Blockchain technologies. The objective of this upgrade was to facilitate more widespread adoption of the technology (Di Francesco Maesa & Mori, 2020).

The latest iteration of Blockchain technology, known as Blockchain 3.0, employs a unique consensus protocol distinct from its predecessors. Rather than utilizing "Proof-of-Work," Blockchain 3.0 relies on "Proof-of-Stake (PoS)"(Sheikh et al., 2018) and "Proof of Authority (PoA)"(Liu et al., 2019) as its consensus protocol (Zhang et al., 2020). These new protocols offer significant advantages for decentralized applications, including scalability, operability, resource efficiency, and a viable alternative to PoW. Consequently, the adoption of Blockchain 3.0 is rapidly gaining traction across various industries (Table 8.1) (Singh & Vadi, 2022).

Table 8.1 Different blockchain generations

	Blockchain 1.0	Blockchain 2.0	Blockchain 3.0	Blockchain 4.0
Application	Monetary region	Non-monetary region	Commerce platforms	Industry 4.0
Instance	Bitcoin	Ethereum	Cardone, IOTA, Anion	Unibright, SEELE
Energy utilization	Maximum	Reasonable	Power competent	Very well-organized
Price	Costly	Inexpensive	Cheaper	Cost-efficient
Speed	7 TBS	15 TBS	1000 TBS	1 M TBS
Scalability	No	Poor	Scalable	Highly scalable
Verification	Via miners	Via smart contracts	In-built confirmation Via dApps	Sharding Automated Confirmation
Consensus mechanism	PoW	Assigned PoW	Proof-of-Stake, authority	Proof of integrity
Interoperability	Not interoperable	Not interoperable	Interoperable	Highly interoperable
Intercommunication	Not allowed	Not allowed	Allowed	Allowed
Primary theory	Distributed Ledger	Smart contracts	dApps	Blockchain with AI
Hashing technique	SHA-256	Ethash	Curl	
Block validation time	10 min	15 s	No blocks	
Chain structure	Meta chain	Meta chain	(a) Meta chain and side chain, (b) Directed graph data structured.	
Scripting language	Very limited	Fully featured programming language	Fully featured programming language	
Features	Guaranteed transaction; authenticity; reduced server costs; transaction transparency	Guarantee of distributed computation; creating and transferring digital assets	Completely open-source; autonomous operation; arbitrary protocol language support	Founded on neural networks and incorporates AI to provide superior decision-making capabilities; improved efficiency; ease of use; top-notch performance
Limitations	Wasted energy and lack of network scalability; Vulnerable to attack and less protected.	Lack of scalability; interoperability, privacy, sustainability, and governance, vulnerable to attack.	Quantum resistant	Still under development and improvement.

Source: Hamdi (2023), Singh and Vadi (2022), and Saxena et al., (2021).

Third-generation Blockchain platforms like Cardano, EOSIO, Zilliqa, ArcBlock, and AION leverage another consensus protocol called Directional Acrylic Graph (DAG) to enhance efficiency, speed, and cost-effectiveness (Kondru & Saranya, 2019). As a result, decentralized applications supported by Blockchain are gaining widespread adoption. Unlike traditional Blockchain technologies, where blocks are linked in a singular chain, DAG allows transactions to connect to multiple previous transactions, enabling faster and more efficient transaction confirmation. DAG-based systems can process numerous transactions simultaneously, avoiding congestion, and transactions can be confirmed as soon as they are approved by other transactions on the network, eliminating the need to wait for a miner to incorporate them into a block (Kotilevets et al., 2018). Since DAG-based systems don't rely on a single chain of blocks, miners don't have to compete to add blocks to the chain, reducing the likelihood of forks and computational power required to confirm transactions. DAG-based systems offer lower fees and faster transaction times than conventional Blockchain systems (Pervez et al., 2018).

8.2.1.4 Blockchain 4.0

In recent studies, Singh and Vadi (2022) and fellow researchers have unveiled the latest iteration of technology. Their findings suggest that the fourth generation of Blockchain technology has the potential to revolutionize various industries. As noted by Singh and Vadi (2022), this new iteration boasts the ability to transcend protocol boundaries and is aided by artificial intelligence. The primary objective of Blockchain 4.0 is to offer a platform for generating and executing applications, ultimately transforming the industry into a fully integrated and adaptable system (Singh & Vadi, 2022).

The most recent iteration of Blockchain technology utilizes advanced consensus algorithms to enhance fault tolerance. This state-of-the-art technology is founded on neural networks and incorporates AI to provide superior decision-making capabilities. Experts predict that it will center on Blockchain-as-a-service (BaaS) and provide improved efficiency, ease of use, and top-notch performance. However, its impact on the industry remains to be thoroughly investigated (Hamdi et al., 2023).

8.2.2 Types of blockchain

There are three main types of blockchain technology: public, private, and consortium, which can be used based on the needs and requirements of the parties involved.

8.2.2.1 Public blockchain

The public blockchain is a decentralized technology, primarily utilized in cryptocurrencies such as Bitcoin. It effectively resolves the issues associated with centralized systems, offering an environment that is both secure and open while also ensuring anonymity (Bhushan et al., 2022). Data is distributed across a peer-to-peer network, and its decentralized nature necessitates an approach for verifying data authenticity. With no central administrator in place, decentralized consensus algorithms are used to make decisions. Notable examples of public blockchains include Litecoin and Bitcoin (Namasudra et al., 2021) (Table 8.2).

8.2.2.2 Private blockchain

A private blockchain is a controlled network that can only be accessed by approved individuals or groups. It is owned by an individual or organization and is not completely decentralized. It provides exceptional efficiency, privacy, and stability. The consensus algorithm is designed with a central authority who has the power to grant mining privileges to interested parties. Private blockchains are both secure and economical. Bankchain is a prime example of a private blockchain (Jayson Baucas et al., 2021) (Table 8.2) and adopting blockchain in supply chain (Maroun & Daniel, 2019; Maroun et al., 2019; Daniel & Maroun, 2022).

8.2.2.3 Consortium blockchain

Consortium blockchains are a hybrid of private and public blockchains, offering faster speed, scalability, and reduced transaction costs compared to private blockchains. Managed by multiple organizations, these semi-decentralized networks involve collaborative decision-making, which enhances their speed while increasing the likelihood of multiple points of failure. Some examples of consortium blockchains include r3 and EWF (Cui et al., 2020).

Table 8.2 Difference between public and private blockchains

Public Blockchain	Private Blockchain
• Anyone can join the public blockchain network	• Only authorized users can join this blockchain network
• Transaction cost is high	• Transaction cost is less
• Transaction speed is high	• Transaction speed is less
• Example: Bitcoin	• Example: Bankchain

Source: Namasudra et al., (2021).

8.2.3 Performance measurement metrics for blockchain technologies

8.2.3.1 Core metrics

Organizations must consider four core metrics when contemplating integrating Blockchain technology into business processes. These metrics provide a valuable tool for assessing the advantages and disadvantages of Blockchain and determining whether it aligns with the organization's needs. A thorough evaluation of these metrics is crucial for a seamless integration of Blockchain technology and a successful outcome.

8.2.3.1.1 Throughput

In blockchain technology, throughput is the count of transactions effectively executed and recorded on the ledger within a defined period. The network's ability to handle traffic determines the throughput, which requires a robust infrastructure. Generally, as the input load grows, the throughput should increase until it reaches the maximum network capacity. Nonetheless, network capacity depends not only on physical resources but also on internal logic execution and access control mechanisms. Robust throughput is a critical requirement for decentralized applications, and it is vital that the system can accommodate a significant number of requests during high-traffic periods within a business environment (Herwanto et al., 2021).

Each Blockchain system has a unique speed for deploying, invoking, and executing while used in the transactions. This makes it crucial to keep track of the system's throughput. Throughput in Blockchain is measured through number of transactions that can be processed by the system per second. By monitoring the throughput, we can evaluate the efficiency and effectiveness of a blockchain system (Zheng et al., 2018).

8.2.3.1.2 Latency

Latency, as it pertains to blockchain, is the amount of time it takes for a transaction to be processed and confirmed on the blockchain network. This duration is the time between when a transaction is submitted and when it is completed. Latency is influenced by various factors, such as propagation delay, transit, and queuing of the system. Additionally, the protocol flow of the system is a significant factor in either increasing or decreasing latency. High latency is a prevalent issue with most established blockchain platforms, such as Bitcoin, due to the size of their ledger. Many businesses are turning to private blockchain systems to avoid additional overheads. The system should be able to respond rapidly, even under high traffic, while simultaneously keeping up with all protocols. Lower latency is typically preferred to ensure faster transaction processing and improve the user experience. High latency can cause delays in transaction confirmation, which reduces the overall speed and efficiency of the blockchain network (Yasaweerasinghelage et al., 2017).

8.2.3.1.3 Resource consumption

The consumption of blockchain resources involves utilizing computational power, storage capacity, bandwidth, and energy by network participants to keep the blockchain operational. Consensus mechanisms such as proof-of-work (PoW) or proof-of-stake (PoS) rely on computational power to verify transactions. As the blockchain ledger grows, storage capacity increases, which requires nodes to store the entire ledger. Bandwidth is necessary for transmitting transactions and blocks across the network. Energy consumption is a significant concern, particularly in PoW systems, because of the intensive mining process. To optimize resource consumption, various efforts are underway, including developing energy-efficient consensus mechanisms, implementing scaling solutions, and advancing hardware technologies. Despite the numerous benefits of blockchain technology, its significant resource requirements remain a primary topic of discussion. Public blockchain systems, such as Bitcoin and Ethereum, are maintained by power and resource-hungry mining rigs, leading businesses to seek private blockchain systems that efficiently utilize resources (Juričić et al., 2020).

8.2.3.1.4 Finality

The term "finality" in blockchain refers to the unchangeable and permanent nature of a transaction once it has been confirmed and added to the blockchain. This feature ensures the security and integrity of transactions because it guarantees that they are recorded permanently on the blockchain ledger. The degree of finality varies depending on the consensus mechanism employed by the blockchain network. In proof-of-work (PoW) blockchains like Bitcoin, the finality of a transaction increases with each additional block added to the chain. Conversely, proof-of-stake (PoS) and other consensus mechanisms can achieve finality more quickly, often with just one confirmation, making transactions more efficient. In summary, finality is a critical component of blockchain technology that instills trust and confidence in the unalterable and trustworthy nature of recorded transactions (Anceaume et al., 2020).

8.2.3.1.5 Scalability

The issue of blockchain scalability has become increasingly pressing as the number of nodes and transactions on the network continues to grow. This poses a challenge for popular public blockchain applications like Bitcoin and Ethereum, where each node must perform a computational task to validate transactions, requiring significant computational power, bandwidth, and storage space.

Transaction throughput and latency are the key performance metrics, with many of the most widely used public blockchains, including Bitcoin and Ethereum, needing help to achieve satisfactory levels. Although these systems can process between 7 and 20 transactions per second, they are often hampered by significant consensus processing time delays, which can be as long as 10 minutes. Moreover, the sheer size of Bitcoin, Ethereum, and Litecoin – 305.23 GB, 667.10 GB, and 28.45 GB, respectively – places considerable demands on storage space and requires significant time for blockchain download (Figure 8.1).

The scalability trilemma initially described by Vitalik Buterin emphasizes the inevitable trade-offs between decentralization, scalability, and security in blockchain technology. This presents a challenge in finding the right balance between these characteristics to meet the needs of public blockchain applications. The focus now is on addressing scalability while maintaining decentralization and security, as this is critical to the future of blockchain technology (Khan et al., 2021).

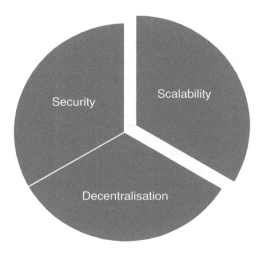

Figure 8.1 Scalability challenges.

Table 8.3 Types of consensus protocol and their fault tolerance capacity

	PoW	PoS	DPoS	PBFT	Ripple
Type	Probabilistic finality	Probabilistic finality	Probabilistic finality	Absolute finality	Absolute finality
Fault tolerance	50%	50%	50%	33%	20%
Power consumption	Large	Less	Less	Negligible	Negligible
Scalability	Good	Good	Good	Bad	Good
Application	Public	Public	Public	Permissioned	Permissioned

Source: Zhang & Lee (2020).

8.2.3.1.6 Fault tolerance – consensus protocol

In Blockchain, various protocols, such as Proof-of-Work (PoW), Proof-of-Stake (PoS), and Delegated Proof-of-Stake (DPoS), utilize probabilistic finality, which necessitates attackers acquiring substantial computational power or stake to substitute a valid chain with a lengthier private chain. In the case of Bitcoin, even a small amount of computational power can lead to a successful double-spend attack, potentially undermining the blockchain network. Similarly, PoS and DPoS protocols only permit stakeholders with holdings below a specific proportion. In Practical Byzantine Fault Tolerance (PBFT), the network must have more normal nodes than a specified threshold, demonstrating the system's fault tolerance of 1/3. Conversely, Ripple can accommodate Byzantine issues with only a fraction of nodes without affecting the consensus outcome (Zhang & Lee, 2020) (Table 8.3).

8.2.4 Blockchain adoption frameworks

8.2.4.1 Technological–organization–environmental framework

Lustenberger et al. (2021) recommend using the Technological–Organization–Environmental (TOE) framework to evaluate a company's readiness to adopt blockchain technology (Lustenberger et al., 2021). This framework considers three critical factors: technological, organizational, and environmental (Depietro et al., 1990). Lustenberger et al., (2021) argues that the TOE framework is particularly useful for investigating the factors that impact blockchain adoption in organizations. The authors also examine several factors that can influence adoption, such as relative advantage, compatibility, complexity, trialability, and observability. Additionally, organizational and environmental factors can impact enthusiasm toward adoption. Finally, "blockchain knowledge" is introduced as a control variable, which differs from the other constructs as it is a personal, not company-related, factor.

8.2.4.2 Blockchain readiness index

Iosif et al., (2021) proposed a blockchain readiness index to assess indicators before adopting blockchain technology. The index compiles various indicators from domains suitable for adopting blockchain technology in a country. The indicators are grouped into technology enablers, industry integration, and governance-related indicators. The index also considers research and user engagement indicators to determine each country's ongoing research efforts and community involvement in blockchain technology and cryptocurrency (Iosif et al., 2021).

8.2.4.3 Other readiness frameworks from the literature

In 2023, Al-Ashmori surveyed software firms to develop a comprehensive framework for blockchain adoption. The framework is based on several critical factors essential for successful adoption, including security, complexity, cost, innovativeness, facilitating conditions, market dynamics, regulatory support, and partner readiness. The survey found that trialability, which involves limited testing before full deployment, increases the likelihood of successful adoption. These factors can also serve as a guiding framework for assessing blockchain technology implementation across various industries (AL-Ashmori et al., 2023).

Balasubramanian et al., (2021) studied organizational factors in the healthcare industry and identified four dimensions of readiness for blockchain implementation – motivational, engagement, technological, and structural (Balasubramanian et al., 2021). Nicolai et al., (2023) recommended using these dimensions to evaluate stakeholder involvement and decision-making in blockchain adoption. Understanding the relationship between these readiness types is crucial for identifying key stakeholders and supporting their decision-making in adopting new processes or technologies, particularly in healthcare where stakeholder acceptance is important (Nicolai et al., 2023).

8.3 MANAGING AND PLANNING BLOCKCHAIN IMPLEMENTATION FRAMEWORK

Organizations must consider several factors before implementing blockchain technologies into their business processes. Figure 8.2 provides a comprehensive framework for managing and planning the integration of blockchain technology in organizations. This framework was developed through an extensive literature review and expert discussions.

8.3.1 Ideation

The process of implementing blockchain technology begins with the crucial ideation phase. In this phase, companies must address three key questions.

Managing and planning blockchain implementation framework 165

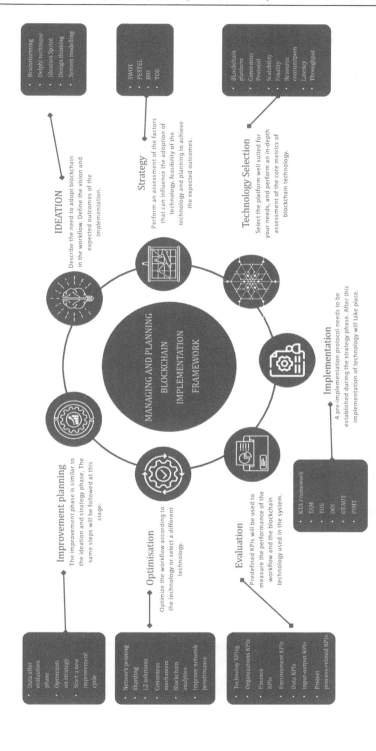

Figure 8.2 Managing and planning blockchain implementation framework.

The first question is why they intend to utilize blockchain technology. The second query involves identifying how the incorporation of blockchain technology will affect their business operations. Lastly, firms must determine what specific results they aim to attain from implementing blockchain technology. The following approaches can be used to answer these pivotal inquiries.

1. Brainstorming – Brainstorming is a problem-solving method where a group spontaneously contributes creative ideas and solutions (Jimoh et al., 2019).
2. Delphi technique – The Delphi method is used to reach a group consensus or decision by gathering opinions from a panel of experts. The experts participate in multiple rounds of questionnaires, and their responses are collected, analyzed, and shared with the group after each round (Rejeb et al., 2022).
3. Ideation Sprint – The Ideation Sprint workshop is a rapid and dynamic creative process that aims to generate validated strategies or concepts to guide your product decisions and directions (Froehlich et al., 2022).
4. Design thinking – Design thinking is a problem-solving approach that puts people at the center. It prompts organizations to prioritize the people they are creating for, resulting in better products, services, and processes. When tackling a business need, the first question should always be: what is the underlying human need we are trying to address? (Schönhals et al., 2018).
5. System modeling – System modeling involves creating abstract models of a system, each providing a unique perspective of the system (Nanda & Nanda, 2022).

These approaches help the organization provide in-depth potential solutions for the problem they are trying to solve by adopting blockchain technology.

8.3.2 Strategy

During the ideation phase, organizations establish clear objectives and potential outcomes. The subsequent strategy phase is a critical step in adopting blockchain technology. As technology constantly evolves, the various available platforms present distinct opportunities and challenges. To achieve success in adopting blockchain technology, organizations must align their chosen platform with their technology adoption objectives and expected outcomes.

The strategy phase is a crucial part of the framework for organizations adopting blockchain technology in their business processes. This phase involves self-reflection and analysis of the organization's capabilities, infrastructure, available digitally literate workforce, work culture, regulations, and environmental factors. SWOT and PESTEL are useful tools for performing these assessments (Nadezda & Josef, 2021).

Before adopting blockchain technology, organizations should also conduct a change readiness assessment of their internal factors. This includes evaluating the motivation of leadership and staff, the digital literacy of current staff, and staff confidence in their leadership's ability to successfully manage the transformation process. Section 8.2.4.3 highlights several frameworks that can prove quite beneficial. Additionally, the balanced scorecard serves as a valuable instrument for self-reflection. It empowers organizations to amalgamate data on service and quality, along with financial performance, into a single report, enhancing operational efficiencies (Govindan et al., 2023).

8.3.3 Technology selection

When an organization incorporates blockchain technology, it is crucial to carefully align its business strategy with its desired outcomes, available resources, and other factors that could influence technology selection. During the technology selection phase, organizations must make informed decisions as they consider the various blockchain platforms that best suit their needs and are compatible with their organization's macro-environmental and internal factors.

Organizations must evaluate different metrics in detail to select the most appropriate blockchain technology, as outlined in Section 8.2.3. Based on these metrics, they can choose the best platform that meets their requirements and can be supported by their available resources and environmental factors. Therefore, it's essential for organizations to carefully assess their actual needs, perform resource analysis, and use those assessments to guide their technology selection process.

8.3.4 Implementation

The next step in the framework involves implementing blockchain technology into business processes. During this phase, organizations must reflect on the assessments conducted in the strategy phase. This reflection helps organizations prepare for implementing blockchain platforms through pre-implementation protocols. These protocols include training and upskilling staff and hiring individuals with higher digital literacy levels (Kardi et al., 2023). Additionally, organizations should invest in more advanced infrastructure and take steps to ensure data synchronization and interoperability between systems. These measures facilitate the mining and validation process of blockchain technologies (Biswas et al., 2020).

Several frameworks have been suggested for incorporating blockchain technology into business processes. Taherdoost (2022) reviewed the literature and proposed Technology Acceptance Model (TAM), TOE framework, Diffusion of Innovations (DOI) framework, and Unified Theory of Acceptance and Use of Technology (UTAUT) framework (Taherdoost, 2022). Janssen et al., (2020) proposed the Process, Institutional, Market, Technology (PIMT) framework, which centers on change strategies and instruments and indicates that institutional, market, and technical factors can influence them (Janssen et al., 2020). Vu et al., (2023) suggested a three-step process for blockchain implementation in food supply chain management, including initiation, adoption decision, and implementation, which can be influenced by innovation, organization, environmental, and manager's characteristics (Vu et al., 2023). Additionally, this chapter suggests using Knowledge to Action (KTA) framework, borrowed from implementation sciences, which can apply knowledge to action in various contexts, but it does not specify the necessary actions after each step, like the frameworks mentioned earlier. (Rycroft-Malone & Bucknall, 2010).

8.3.5 Evaluation

In the evaluation phase, organizations must revisit the objectives and anticipated outcomes of their blockchain technology implementation, as established during the ideation phase of the framework. The second step in this phase involves reflecting on the organization's self-assessment during the strategy phase, which is essential for analyzing trade-offs. Organizations need to consider a range of key performance indicators (KPIs) to assess whether blockchain technology is achieving its intended goals. These include factors such as cost, time, speed, dependability, efficiency, process effectiveness, scalability, flexibility, resilience, security, transparency, and sustainability. Depending on the organization and platforms used, KPIs can be defined and performance of blockchain technology in practice can be assessed (Raval et al., 2022). Therefore, it is essential to gather information during the ideation and strategy phases so that organizations can evaluate these KPIs effectively.

8.3.6 Optimization

The optimization phase follows if an organization's blockchain implementation still needs to meet its intended objectives. Fortunately, advancements in blockchain have yielded strategies that can improve network performance and scalability. For example, network pruning involves systematically removing obsolete and redundant data from the blockchain to improve efficiency (Palm et al., 2018). Sharding consists of partitioning the blockchain into smaller, more manageable segments, allowing for parallel transaction processing and increasing transaction throughput (Hashim et al., 2022). Layer 2 (L2) solutions provide a complementary strategy by offloading transactions from the main net to a secondary layer, reducing the burden on the main net and enabling it to operate more efficiently and at a lower cost (Xu & Chen, 2022). Additionally, consensus mechanisms have evolved to include more efficient algorithms that improve network throughput and transaction finality, mitigating the limitations of earlier systems. (Yadav et al., 2023).

Furthermore, effective resource management is critical to ensure optimal network operations, which requires advanced tools to monitor resource usage. The backbone of blockchain maintenance lies in monitoring and analysis, leveraging analytics tools to provide real-time insights and in-depth historical data analysis (Balaskas & Franqueira, 2018). Evaluating a blockchain's performance metrics through benchmarking against similar networks is crucial to gain a clear perspective of its relative strengths and weaknesses (Wang et al., 2019). Finally, stress testing is vital in identifying potential network issues by simulating high-load scenarios to uncover potential bottlenecks and guide optimization efforts. Finally, in the rapidly evolving blockchain landscape, it is essential to continuously evaluate and adapt by regularly assessing network performance and integrating innovative strategies to maintain and enhance the scalability and efficacy of blockchain systems (Antwi et al., 2022).

8.3.7 Improvement planning

The framework's ultimate phase entails improvement planning. Achieving optimal blockchain technology may necessitate modifying the organization's infrastructure or work culture. Hence, this improvement planning phase aligns with the ideation phase, where novel optimization strategies are deliberated. Subsequently, the strategy phase commences, wherein the planning to optimize business processes is executed. This cycle persists until the desired objectives are met.

8.4 DISCUSSION

The ideation phase is pivotal to digital transformation management, change management, and enhancing business processes, products, or services. During this phase, the organization identifies its requirements and evaluates all possible opportunities these ideas can offer for organizational growth. Multiple approaches have been mentioned in Section 8.3.1, taken from the literature. In this phase, the organization's innovation management team puts all ideas on the table and selects the ones with the highest potential to move forward to the strategy phase. At this stage, the company must define its goals and should be able to answer how blockchain adoption would help them achieve those goals as part of value creation.

Once objectives and expected outcomes have been established, the next step is to develop strategies for implementing blockchain technologies. This will help achieve the objectives and create value for the organization. Organizations must consider external macro-environmental and internal organizational factors during the strategy phase before selecting the most appropriate blockchain technologies to meet their needs. Nadezda and Josef (2021) suggest using SWOT analysis to assess macro-environmental factors, while Section 8.2.4.3 covers multiple frameworks for assessing organizational readiness for blockchain technology adoption. Organizations can select the best-suited blockchain technology for their needs by conducting these two assessments.

When considering which blockchain platform to adopt, organizations must reflect on their predefined objectives for using blockchain and the intended outcomes they hope to achieve. Organizations can refer back to their self-assessment from the strategy phase to make informed decisions during the technology selection phase. Khan et al., (2021) recommended using the scalability trilemma presented by Vitalik Buterin. This trilemma outlines the trade-offs between decentralization, scalability, and security, allowing organizations to prioritize their needs and select the most suitable blockchain platform. However, choosing one purpose from the trilemma means making compromises on the other two purposes of blockchain technology. Therefore, evaluating the technology's performance after implementation is crucial and using a loop mechanism to improve the business processes is vital.

For organizations looking to adopt blockchain technology, it's essential to follow pre-implementation protocols to ensure readiness. While various frameworks are available for the implementation phase, many are centered around assessment strategies rather than the actual implementation itself. Taherdoost (2022) suggests using adoption frameworks not specific to blockchain technology to evaluate implementation outcomes best. Additionally, while Janssen's et al., (2020) PIMT framework does not provide an in-depth analysis of the technology's challenges, it does consider external factors that can impact the process. Vu et al., (2023) recommends a four-step, industry-specific framework. Still, it is worth noting that implementing blockchain technology is a complex process requiring thorough assessment before adoption and evaluation of the technology's performance after implementation.

Various blockchain frameworks were discussed earlier, but no evaluation criteria for measuring blockchain implementation were mentioned. Zhang et al., (2020) proposed an industry-specific framework that suggested considering product design, development, and strategic planning for improvement after implementation. Unfortunately, the framework did not provide Key Performance Indicators (KPIs) for evaluating system performance after implementation (Zhang et al., 2020). However, Raval et al., (2022) proposed a set of KPIs that can be used to measure blockchain technology performance in different industries. Although Raval's work is industry-specific, most of the KPIs defined in the study can be applied to other sectors. The KPIs were categorized into eight categories, seven of which were transferable to other industries. Raval highlighted several significant KPIs, including technology, organizations, finance, environment, data management and security, input–output, and project process-related KPIs.

Organizations must utilize Key Performance Indicators (KPIs) in their optimization process to achieve their objectives in blockchain implementation initiatives. Section 3.6 delves into various optimization strategies that are particularly beneficial during the improvement phase. While blockchain technology is still evolving, research is ongoing to optimize its utilization, and new strategies will emerge. For now, the most suitable strategy must be narrowed down, and the most appropriate one should be selected for the improvement planning phase. This blockchain management and implementation cycle will continue until organizations realize their intended blockchain implementation objectives.

8.4.1 Practical contribution

The blockchain management and implementation framework proposed in this chapter offers organizations a comprehensive guide for implementing blockchain technology into their business processes. It provides strategic assessment guidelines, frameworks for implementation, KPIs for performance evaluation, and optimization strategies. These contributions are carefully designed to facilitate each phase of blockchain implementation, from ideation to continuous improvement. By navigating the complexities of blockchain technology with this guide, organizations can realize its full potential and reap its benefits.

8.4.2 Academic contribution

Previous research has proposed various approaches for implementing blockchain technology. Some have recommended leveraging conventional technology adoption frameworks, while others have emphasized adopting industry-specific blockchain implementation frameworks. However, these frameworks need to emphasize conducting a comprehensive analysis of blockchain platforms before implementation or performing evaluations of blockchain-supported business processes. The chapter's framework for managing and implementing blockchain highlights the importance of thoroughly analyzing available blockchain platforms and considering performance metrics. Furthermore, it suggests optimizing and evaluating the technology's performance in real-world scenarios to achieve desired objectives, filling a significant gap in existing literature.

8.5 CONCLUSION AND FUTURE DIRECTIONS

Blockchain technology is advancing rapidly, and ongoing research is being conducted to improve its functionality. It is undoubtedly a wise decision for any organization to adopt this technology for value creation. Still, planning thoroughly and remaining open to further improvements is crucial before implementing blockchain into business processes. This chapter proposes a comprehensive framework that guides organizations seeking to implement blockchain technology successfully. However, since the framework has yet to be utilized, further research must validate its effectiveness or identify any necessary improvements. Organizations are encouraged to use this framework and share their findings through case reports.

REFERENCES

AL-Ashmori, A. et al. (2023) 'A readiness model and factors influencing blockchain adoption in Malaysia's software sector: A survey study'. *Sustainability*, 15(16), p. 12139.

Anandan, R. and Deepak, B.S. (2020) 'An overview of blockchain technology: fundamental theories and concepts'. In *The Convergence of Artificial Intelligence and Blockchain Technologies*. World Scientific, pp. 1–22. doi: 10.1142/9789811225079_0001.

Anceaume, E. et al. (2020) (arXiv:2012.10172) Available at: https://arxiv.org/abs/2012.10172 (Accessed: 15 February 2024).

Antwi, R. et al. (2022) (6) 'A survey on network optimization techniques for blockchain systems'. *Algorithms*, 15(6), p. 193. doi:10.3390/a15060193.

Auer, R. (2019) 'Beyond the Doomsday economics of 'proof-of-work' in cryptocurrencies'. p. 31. BIS Working Papers 765, Bank for International Settlements.

Balaskas, A. and Franqueira, V.N.L. (2018) 'Analytical tools for blockchain: Review, taxonomy and open challenges'. In *2018 International Conference on Cyber Security and Protection of Digital Services (Cyber Security)*. 2018 International Conference on Cyber Security and Protection of Digital Services (Cyber Security). pp. 1–8. doi: 10.1109/CyberSecPODS.2018.8560672.

Balasubramanian, S. et al. (2021) 'A readiness assessment framework for blockchain adoption: A healthcare case study'. *Technological Forecasting and Social Change*, 165, p. 120536.

Bhushan, B. et al. (2022) 'Leveraging blockchain technology in sustainable supply chain management and logistics'. *Blockchain Technologies for Sustainability*, 1st ed. 2022, pp. 179–196.

Biswas, S. et al. (2020) 'Interoperability and synchronization management of blockchain-based decentralized e-health systems'. *IEEE Transactions on Engineering Management*, 67(4), pp. 1363–1376.

Böhme, R. et al. (2015) 'Bitcoin: Economics, technology, and governance'. *Journal of Economic Perspectives*, 29(2), pp. 213–238.

Buterin, V. (2013) 'Ethereum White Paper'. *GitHub Repository*, 1, pp. 22–23.

Buterin, V. (2022) *Proof of Stake: The Making of Ethereum and the Philosophy of Blockchains*. New York, United States: Seven Stories Press. ISBN 164421248X, 9781644212486

Chen, K. et al. (2023) 'Research on the application of penetration testing frameworks in blockchain security'. In *International Conference on Computational & Experimental Engineering and Sciences*. Springer Nature Switzerland, pp. 307–330. Proceedings of ICCES 2023—Volume 3, Shenzhen China.

Chen, T. et al. (2017) 'Under-Optimized Smart Contracts Devour Your Money'. In *2017 IEEE 24th International Conference on Software Analysis, Evolution and Reengineering (SANER)*. IEEE, Klagenfurt, Austria, pp. 442–446. doi: 10.1109/SANER.2017.7884650

Cui, Z. et al. (2020) 'A hybrid blockchain-based identity authentication scheme for multi-WSN'. *IEEE Transactions on Services Computing*, 13(2), pp. 241–251.

Dabbagh, M., Sookhak, M. and Safa, N.S. (2019) 'The evolution of blockchain: A bibliometric study'. *IEEE Access*, 7, pp. 19212–19221. DOI: 10.1109/ACCESS.2019.2895646.

Daniel, J., & Maroun, E. (2022). Blockchain and sustainability: Can blockchain technology improve supply chain sustainability? In *International Conference of the Production and Operations Management Society* (POMS).

De Vries, A. (2018) 'Bitcoin's Growing Energy Problem'. *Joule*, 2(5), pp. 801–805.

Delmolino, K. et al. (2016) 'Step by Step towards Creating a Safe Smart Contract: Lessons and Insights from a Cryptocurrency Lab'. In *Financial Cryptography and Data Security: FC 2016 International Workshops, BITCOIN, VOTING, and WAHC, Christ Church, Barbados,* February 26, 2016, Revised Selected Papers 20. Springer, pp. 79–94.

Depietro, R., Wiarda, E. and Fleischer, M. (1990) 'The context for change: Organization, technology and environment'. *The Processes of Technological Innovation*, 199, pp. 151–175.

Di Francesco Maesa, D. and Mori, P. (2020) 'Blockchain 3.0 Applications Survey'. *Journal of Parallel and Distributed Computing*, 138, pp. 99–114. doi: 10.1016/j.jpdc.2019.12.019.

Diffie, W. and Hellman, M.E. (2022) 'New directions in cryptography'. In *Democratizing Cryptography: The Work of Whitfield Diffie and Martin Hellman.* pp. 365–390. Association for Computing Machinery, New York NY United States. ISBN:978-1-4503-9827-5. doi: https://doi.org/10.1145/3549993

Dumitrescu, G.C. (2017) 'Bitcoin–a Brief Analysis of the Advantages and Disadvantages'. *Global Economic Observer*, 5(2), pp. 63–71.

Froehlich, M. et al. (2022) 'Prototyping with blockchain: A case study for teaching blockchain application development at University'. In *International Conference on Interactive Collaborative Learning*. Vienna, Austria: Springer, pp. 1005–1017. Lecture Notes in Networks and Systems, ISSN 2367-3389.

Govindan, K. et al. (2023) 'Prioritizing adoption barriers of platforms based on blockchain technology from balanced scorecard perspectives in healthcare industry: A structural approach'. *International Journal of Production Research*, 61(11), pp. 3512–3526.

Haber, S. and Stornetta, W.S. (1991) *How to Time-Stamp a Digital Document.* New Jersey, United States: Springer. *J. Cryptology*, 3, 99–111. https://doi.org/10.1007/BF00196791

Hamdi, A., Fourati, L. and Ayed, S. (2023) 'Vulnerabilities and attacks assessments in blockchain 1.0, 2.0 and 3.0: Tools, analysis and countermeasures'. *International Journal of Information Security*, pp. 1–45. doi: 10.1007/s10207-023-00765-0.

Hashim, F., Shuaib, K. and Zaki, N. (2022) 'Sharding for scalable blockchain networks'. *SN Computer Science*, 4(1), p. 2. doi: 10.1007/s42979-022-01435-z.

Herwanto, R., Sabita, H. and Armawan, F. (2021) 'Measuring throughput and latency distributed ledger technology: Hyperledger'. *Journal of Information Technology Ampera*, 2(1), pp. 17–31.

Iosif, E., Christodoulou, K. and Vlachos, A. (2021) 'A robust blockchain readiness index model'. *arXiv Preprint arXiv:2101.09162*.

Janssen, M. et al. (2020) 'A framework for analysing blockchain technology adoption: Integrating institutional, market and technical factors'. *International Journal of Information Management*, 50, pp. 302–309.

Jayson Baucas, M., Gadsden, S.A. and Spachos, P. (2021) 'IoT-based smart home device monitor using private blockchain technology and localization'. *arXiv E-Prints*, p. arXiv–2103.

Jena, A.K. and Dash, S.P. (2021) 'Blockchain technology: Introduction, applications, challenges'. In Panda, S. K. et al. (eds.) *Blockchain Technology: Applications and Challenges*. Cham: Springer International Publishing, pp. 1–11. doi: 10.1007/978-3-030-69395-4_1.

Jimoh, F. O., Abdullahi, U. G. and Ibrahim, I. A. (2019) 'An overview of blockchain technology adoption'. *Journal of Computer Science and Information Technology*, 7(2). doi: 10.15640/jcsit.v7n2a4.

Johnston, D. et al. (2014) 'The General Theory of Decen-tralized Applications, DApps'. Technology and Investment, 12(3).

Juričić, V., Radošević, M. and Fuzul, E. (2020) 'Optimizing the resource consumption of blockchain technology in business systems'. *Business Systems Research: International Journal of the Society for Advancing Innovation and Research in Economy*, 11(3), pp. 78–92.

Kardi, K. et al. (2023) 'The nexus of artificial intelligence, blockchain technology, and human capital in digital marketing strategy: An exploratory study on the integration, ethical implications, and future prospects'. *International Journal of Economic Literature*, 1(1), pp. 12–22.

Khan, D., Jung, L.T. and Hashmani, M.A. (2021) (20) 'Systematic literature review of challenges in blockchain scalability'. *Applied Sciences*, 11(20), p. 9372. doi: 10.3390/app11209372.

Kondru, K.K. and Saranya, R. (2019) 'Directed acyclic graph-based distributed ledgers—an evolutionary perspective'. *Int J Eng Adv Technol (IJEAT)*, 9(1), 6096–6103

Kotilevets, I. et al. (2018) 'Implementation of directed acyclic graph in blockchain network to improve security and speed of transactions'. *IFAC-PapersOnLine*, 51(30), pp. 693–696.

Lee, J.Y. (2019) 'A decentralized token economy: How blockchain and cryptocurrency can revolutionize business'. *Digital Transformation & Disruption*, 62(6), pp. 773–784. doi: 10.1016/j.bushor.2019.08.003.

Liu, X. et al. (2019) 'MDP-based quantitative analysis framework for proof of authority'. In *2019 International Conference on Cyber-Enabled Distributed Computing and Knowledge Discovery (CyberC)*. IEEE, pp. 227–236.

Lustenberger, M., Malešević, S. and Spychiger, F. (2021) 'Ecosystem readiness: Blockchain adoption is driven externally'. *Frontiers in Blockchain*, 4, p. 720454.

Malik, N. et al. (2022) 'Why bitcoin will fail to scale?' *Management Science*, 68(10), pp. 7323–7349. doi: 10.1287/mnsc.2021.4271.

Maroun, E. A. and Daniel, J. (2019). Opportunities for use of blockchain technology in supply chains: Australian manufacturer case study. *Proceedings of the 9th International Conference on Industrial Engineering and Operations Management*. Bangkok, Thailand. ISBN: 978-1-5323-5948-4

Maroun, E.A., Daniel, J., Zowghi, D., & Talaei-Khoei, A. (2019). Blockchain in supply chain management: Australian manufacturer case study. In *Service Research and Innovation: 7th Australian Symposium, ASSRI 2018*, Sydney, NSW, Australia, September 6, 2018, and Wollongong, NSW, Australia, December 14, 2018, Revised Selected Papers 7 (pp. 93–107). Springer International Publishing.

Mignon, V. (2019) 'Blockchains - perspectives and challenges'. In *Blockchains, Smart Contracts, Decentralised Autonomous Organisations and the Law*. Edward Elgar Publishing, pp. 1–17. Available at: https://www.elgaronline.com/edcollchap/edcoll/9781788115124/9781788115124.00007.xml (Accessed: 7 February 2024).

Mitrofanova, I.A. (2018) 'The legislative regulation of "smart" contracts: The problems and prospects of development'. *Legal Concept= Pravovaya Paradigma*, 17(4).

Nadezda, F. and Josef, A. (2021) 'Economic perspectives of the blockchain technology: Application of a SWOT analysis'. *Terra Economicus*, 19(1), pp. 78–90.

Nakamoto, S. (2008) 'Bitcoin: A Peer-to-Peer Electronic Cash System'. Decentralized Business Review.

Namasudra, S. et al. (2021) 'The Revolution of Blockchain: State-of-the-Art and Research Challenges'. *Archives of Computational Methods in Engineering*, 28(3), pp. 1497–1515. doi: 10.1007/s11831-020-09426-0.

Nanda, S.. and Nanda, S.. (2022) 'Blockchain adoption in health market: A systems thinking and modelling approach'. *Journal of Asia Business Studies*, 16(2), pp. 396–405.

Nicolai, B. et al. (2023) 'Blockchain for electronic medical record: Assessing stakeholders' readiness for successful blockchain adoption in health-care'. *Measuring Business Excellence*, 27(1), pp. 157–171.

Palm, E., Schelen, O. and Bodin, U. (2018) 'Selective blockchain transaction pruning and state derivability'. In *2018 Crypto Valley Conference on Blockchain Technology (CVCBT)*. Zug: IEEE, pp. 31–40. doi: 10.1109/CVCBT.2018.00009.

Pervez, H. et al. (2018) 'A comparative analysis of DAG-based blockchain architectures'. In *2018 12th International Conference on Open Source Systems and Technologies (ICOSST)*. Lahore Pakistan: IEEE, pp. 27–34. doi: 10.1109/ICOSST.2018.8632193

Raval, P., Sarkar, D. and Devani, D. (2022) 'Application of analytical-network-process (ANP) for evaluation of key-performance-indicators (KPI) for application of blockchain technology in infrastructure projects'. *Innovative Infrastructure Solutions*, 7(6), p. 358. doi: 10.1007/s41062-022-00952-3.

Rejeb, A. et al. (2022) (6) 'Barriers to blockchain adoption in the circular economy: A fuzzy Delphi and best-worst approach'. *Sustainability*, 14(6), p. 3611. doi: 10.3390/su14063611.

Rycroft-Malone, J. and Bucknall, T. (2010) *Models and Frameworks for Implementing Evidence-Based Practice: Linking Evidence to Action*. New Jersey, United States: John Wiley & Sons. ISBN: 978-1-444-35873-5

Saraji, S. (2023) 'Introduction to blockchain'. In *Sustainable Oil and Gas Using Blockchain*. New Jersey, United States: Springer, pp. 57–74.

Saxena, R., Arora, D. and Mahapatra, S. (2021) 'Bitcoin: A digital cryptocurrency'. In pp. 13–28. doi: 10.1007/978-3-030-69395-4_2.

Schönhals, A., Hepp, T. and Gipp, B. (2018) 'Design thinking using the blockchain: Enable traceability of intellectual property in problem-solving processes for open innovation'. In *Proceedings of the 1st Workshop on Cryptocurrencies and Blockchains for Distributed Systems.* pp. 105–110. Munich Germany. ISBN: 978-1-4503-5838-5

Sheikh, H., Azmathullah, R.M. and Rizwan, F. (2018) 'Proof-of-work vs proof-of-stake: A comparative analysis and an approach to blockchain consensus mechanism'. *International Journal for Research in Applied Science & Engineering Technology,* 6(12), pp. 786–791.

Singh, D. and Malhotra, M.V. (2023) 'A review on the capability and smart contract potential of block chain technology'. In *2023 3rd International Conference on Smart Data Intelligence (ICSMDI).* Trichy, India: IEEE, pp. 80–87. doi: 10.1109/ICSMDI57622.2023.00022

Singh, G., Garg, V. and Tiwari, P. (2021) 'Introduction to blockchain technology'. In Agrawal, R. and Gupta, N. (eds.) *Transforming Cybersecurity Solutions Using Blockchain.* Singapore: Springer Singapore, pp. 1–18. DOI: 10.1007/978-981-33-6858-3_1.

Singh, S. and Vadi, V. (2022) Evolutionary Transformation of Blockchain Technology. *International Journal of Engineering Research & Technology (IJERT)* ISSN: 2278-0181 IHIC - 2021 Conference Proceedings.

Swan, M. (2015) *Blockchain: Blueprint for a New Economy.* California, United States: O'Reilly Media, Inc ISBN: 1491920475, 9781491920473

Taherdoost, H. (2022) 'A critical review of blockchain acceptance models—blockchain technology adoption frameworks and applications'. *Computers,* 11(2), p. 24.

Vu, N., Ghadge, A. and Bourlakis, M. (2023) 'Blockchain adoption in food supply chains: A review and implementation framework'. *Production Planning & Control,* 34(6), pp. 506–523.

Vukolić, M. (2016) 'The quest for scalable blockchain fabric: proof-of-work vs. BFT replication'. In Camenisch, J. and Kesdoğan, D. (Eds.) *Open Problems in Network Security.* Cham: Springer International Publishing, pp. 112–125.

Wang, R., Ye, K. and Xu, C.-Z. (2019) 'Performance benchmarking and optimization for blockchain systems: A survey'. In Joshi, J. et al. (Eds.) *Blockchain – ICBC 2019.* Cham: Springer International Publishing, pp. 171–185.

Xu, Z. and Chen, L. (2022) 'L2chain: Towards high-performance, confidential and secure layer-2 blockchain solution for decentralized applications'. *Proceedings of the VLDB Endowment,* 16(4), pp. 986–999. doi: 10.14778/3574245.3574278.

Yadav, A.K. et al. (2023) 'A comparative study on consensus mechanism with security threats and future scopes: Blockchain'. *Computer Communications,* 201, pp. 102–115.

Yasaweerasinghelage, R., Staples, M. and Weber, I. (2017) 'Predicting latency of blockchain-based systems using architectural modelling and simulation'. In *2017 IEEE International Conference on Software Architecture (ICSA).* Gothenburg, Sweden: IEEE, pp. 253–256. doi: 10.1109/ICSA.2017.22

Zarir, A.A. et al. (2021) 'Developing cost-effective blockchain-powered applications: A case study of the gas usage of smart contract transactions in the Ethereum blockchain platform'. *ACM Transactions on Software Engineering and Methodology*, 30(3), p. 28:1–28:38. doi: 10.1145/3431726.

Zhang, A. et al. (2020) 'Blockchain-based life cycle assessment: An implementation framework and system architecture'. *Resources, Conservation and Recycling*, 152, p. 104512. DOI: 10.1016/j.resconrec.2019.104512.

Zhang, S. and Lee, J.-H. (2020) 'Analysis of the main consensus protocols of blockchain'. *ICT Express*, 6(2), pp. 93–97. doi: 10.1016/j.icte.2019.08.001.

Zheng, P. et al. (2018) 'A detailed and real-time performance monitoring framework for blockchain systems'. In *2018 IEEE/ACM 40th International Conference on Software Engineering: Software Engineering in Practice Track (ICSE-SEIP)*. pp. 134–143. Available at: https://ieeexplore.ieee.org/document/8449244 (Accessed: 28 February 2024).

Chapter 9

Ensuring transparency

The blockchain impact on agriculture and livestock traceability

Rakesh G Nair, Pankaj Kumar Detwal, and M. Muthukumar

9.1 INTRODUCTION

The livestock and agriculture sectors are the fundamentals of human survival, acting as the principal elements of nutritional well-being and food security. These crucial industries serve as economic drivers, wielding sizable influence on national gross domestic product (GDP) and fostering key avenues of employment globally (Khan et al., 2020; Loizou et al., 2019). Conversely, contemporary agricultural practices are encountering a barrage of challenges. Soil deterioration, flawed cultivation systems, fluctuations in market prices, and erratic climate patterns all affect crop yield and stability (Liu et al., 2015). The consequential effects extend to the interrelated domain of the livestock sector, adversely influencing market confidence and the availability of feed (Bocquier & González-García, 2010).

Adding to the complexity, an increasing global population coupled with limited resources exacerbates the demand for food which in turn leads to broadened supply–demand gap (Bocquier & González-García, 2010; Liu et al., 2015). This further transfers the challenges all along the agricultural supply chain, inducing apprehensions regarding quality assurance, buyer trust, and product traceability amidst the imminent specter of corporate associations (Bosona & Gebresenbet, 2013). The intricate nature of the agricultural supply chain network, containing a large number of stakeholders with often conflicting objectives, further complicates the quest for transparency and traceability from production to consumption (Dabbene et al., 2014; Sarpong, 2014). Farmers, wholesalers, intermediaries, retailers, regulatory entities, and customers interrelate in a complex ecosystem, enforcing the requirement for novel solutions to ensure transparent and ethical food production.

Customers and intermediaries within the supply chain insist on safe and top-quality products with exhaustive information about food attributes, animal welfare standards, country of origin, and transparency regarding genetically modified organisms (Bastian & Zentes, 2013). Additionally, the authentication of food products plays a pivotal role in validating label information, encapsulating aspects such as production methodologies, geographical origin, processing techniques, and constituent composition (Danezis et al., 2016). The unorganized sector within the agri-food industry, inhabited by both small-scale producers and influential corporations, is explicitly susceptible to disparities in information, thereby potentially jeopardizing the interests of susceptible stakeholders (Kumar et al., 2022). According to "Mao et al., (2018) asymmetric information ensues when parties involved in an economic transaction are not equally informed and prevents the first-best allocation of resources causing a market failure. Therefore, the intricate structure of the agri-livestock supply chain presents a significant challenge for integrating data and orchestrating processes across numerous stakeholders while ensuring transparency, efficiency, security, and safety (Kache & Seuring, 2017).

In recent years, blockchain has risen to prominence, dominating discussions in the media, industry-focused publications, and the public unveilings of novel startups. By now, it's likely that most individuals have come across the term "blockchain technology," often in connection with Bitcoin. Nevertheless, it's essential to acknowledge that blockchain's scope reaches far beyond digital currency and the financial sector. Then what exactly blockchain technology is? In simpler terms, blockchain technology is actually a powerful type of secure database that is jointly created by the users in the network (Antonucci et al., 2019). Blockchain is termed a chain of blocks (Niranjanamurthy et al., 2019). Data records are generated within the blockchain, and each successive block is linked to the one that precedes it. (El Koshiry et al., 2023). The most recent block in this sequence encapsulates the entire historical record of the chain, resulting in the formation of the blockchain (Niranjanamurthy et al., 2019).

You may question whether it functions like a database, but what sets it apart from relational databases like Oracle, MS Access, MySQL, and similar systems? What sets blockchain apart is its inherent transparency and immutability? Information stored on blockchain networks is publicly accessible and once recorded, it remains incorruptible, as no data within the blockchain can be tampered with or altered (Minoli & Occhiogrosso, 2018). In simpler terms, once data is entered into the blockchain, it cannot be altered or deleted without the permission of other users and the new blocks of data created will contain the information from previous blocks (Niranjanamurthy et al., 2019). Concisely, we can say that it is a completely decentralized database network.

The concept of blockchain was introduced by an individual or group using the pseudonym "Satoshi Nakamoto" in 2008, with the primary purpose of serving as the public transaction ledger for the cryptocurrency known as Bitcoin (Tredinnick, 2019). Before Bitcoin's groundbreaking emergence, Satoshi Nakamoto envisioned a radical financial system (Tredinnick, 2019). This system would bypass traditional intermediaries, empowering individuals to manage their finances without relying on centralized institutions.

Bitcoin, a cryptocurrency, uses blockchain technology to securely document transactions in a public ledger (Nakamoto, 2008). But blockchain's potential goes far beyond cryptocurrency. This shared ledger, also identified as Distributed Ledger Technology (DLT) (Minoli & Occhiogrosso, 2018), functions as a protected repository for information. Each block on the ledger contains data and transaction particulars and possesses an intrinsic tamper-proof quality (Tredinnick, 2019). This signifies once data is inscribed, it's exceedingly difficult to modify or tamper with, providing an unrivaled degree of trust and transparency (Nakamoto, 2008).

From tracing the journey of a tomato to elucidating the lineage of gold, blockchain' capability of facilitating secure record-keeping revolutionizes diverse businesses, involving livestock and agriculture (Antonucci et al., 2019). Within these complex supply chains, where various stakeholders interact, blockchain bestows control and visibility. Farmers, consumers, and retailers can access and authenticate the data, ensuring shared ownership and accountability. Once authorized, information becomes intrinsic to the blockchain, creating an immutable and verifiable record (Kumar et al., 2022).

For the agri-livestock industry, blockchain presents a transformative possibility. Imagine a single, secure platform which records a farm's entire history, inventory details, its operational status, and even contractual agreements (Cao et al., 2022). Presently, many farmers rely on a mix of software, applications, spreadsheets, manual notes, and memory to document their data (Alshehri, 2023). This complexity multiplies when farm service providers require access to this data for their services. By establishing a singular data source for farms, blockchain reduces the burden of managing multiple record-keeping systems (Alshehri, 2023). Ultimately, blockchain has the potential to save time and streamline operations across the entire value chain. Before we investigate the application of blockchain technology in livestock sector, let's reacquaint with the challenges of livestock sector and how the application of blockchain solutions can alleviate the glitches of the sector. Some of the potential obstacles that push back the meat sector are:

a. Complex/traditional supply chain
b. Diseases and animal welfare issues
c. Traceability issues in the supply chain
d. Unhygienic slaughtering practices
e. Environmental issues
f. Meat safety and quality control
g. Changing food consumption patterns

In spite of the many positive developments and promising applications in the meat sector, the industry is still depending on the traditional supply chain and the players are also using the outdated communication methods to flow information across the chain (Alshehri, 2023; Patel et al., 2023). This creates and widens the bullwhip. Blockchain technology can reduce these uncertainties to an extent (Alshehri, 2023). Blockchain, with its secure distributed ledger and decentralized computing, offers a promising solution to enhance transparency, traceability, security, and trust, ultimately ensuring profitability for all stakeholders in the agricultural and allied ecosystem (Cao et al., 2022). Within this chapter, we will explore in depth the multifaceted application of blockchain technologies and how they are reshaping and revolutionizing the landscape of agriculture and the livestock industry.

9.2 APPLICATION OF BLOCKCHAIN TECHNOLOGY IN THE LIVESTOCK SECTOR

In the livestock sector, the implementation of blockchain solutions holds promise for several specific use cases that can deliver substantial value to the industry's entire supply chain while also addressing some of its key challenges.

9.2.1 Supply chain traceability

In contemporary times, animal identification and traceability play pivotal roles as management tools in ensuring animal health and food safety (Patel et al., 2023). Traceability, in essence, involves the capability to record all pertinent components, encompassing movements, procedures, and requisite controls, to comprehensively outline the life history of an animal or animal product (Smith et al., 2005). The integration of traceability, coupled with routine animal testing, empowers authorities to swiftly pinpoint the origins of potential threats to both animal and human health (Jiang et al., 2023). This, in turn, reduces the risk of broader disease outbreaks and reinforces quality assurance procedures for animals and their products (Jiang et al., 2023). Nonetheless, the traceability systems currently under development in various countries exhibit substantial differences in their specific attributes (Hobbs, 2004). This disparity has led to trade-related concerns in global markets, particularly concerning live cattle and beef products (Sarpong, 2014). Currently, there is no comprehensive international consensus regarding the precise definition of traceability (Patel et al., 2023). Nevertheless, intensive discussions are ongoing within prominent bodies such as the Codex Alimentarius Commission, the World Trade Organization (WTO), and the International Organization for Standardization (ISO) with the aim of reaching a consensus on this matter (Smith et al., 2005).

As far as developing countries like India, the livestock holding size by households is ridiculously small which makes it difficult to establish animal identification and traceability systems (Dandage et al., 2017). However, in poultry sector, the players are well organized than livestock sector and the traceability system can be implemented easily (Dandage et al., 2017). Traceability has evolved into a crucial component of supply chains in recent times, driven by increased consumer awareness and its capacity to enhance planning and issue identification. Within complex supply networks, there is an anticipation that both lead times and costs could suffer negative consequences without adequate traceability (Sarpong, 2014). Furthermore, the bullwhip effect, which disrupts the seamless operation of the supply chain, can be mitigated with the application of blockchain technology (Sarpong, 2014). Blockchain allows for the comprehensive tracking of the entire physical production process, thereby helping to reduce the bullwhip effect. This system operates cohesively across multiple entities, as their progress relies on accurate inputs (Patel et al., 2023). Every stage of the production process maintains a prominent level of accountability, ensuring that traceability is not limited to a single product but extends to encompass its constituent ingredients as well (Alshehri, 2023).

9.2.2 Quality and safety assurance

Quality is all about satisfying stated and implied needs – now or in the future. Despite the presence of regulatory measures, foodborne infections persist as a formidable challenge, resulting in millions of cases on a global scale each year (Motarjemi & Käferstein, 1999). To address this ongoing issue, the food industry must proactively devise and adopt diverse food safety programs, approaches, and technological advancements to mitigate pathogenic bacteria in raw and ready-to-eat (RTE) meat products (Jiang & Xiong, 2015). This encompasses combating pathogens like *E. coli* O157:H7, Salmonella, and Listeria monocytogenes, as highlighted by Qi Zhu in 2017 (Sheng & Zhu, 2021).

Proper and hygienic handling of carcasses post-slaughter holds immense importance in averting contamination and ensuring meat safety, regardless of whether it pertains to formal or informal meat trading sectors (Hauge et al., 2012). This critical aspect, underscored by Warriss, 1990, underscores how the methods employed during slaughter and carcass management directly influence meat quality and its shelf life. Improper animal handling can result in carcass and meat quality problems, including pale soft exudative and dark firm dry meat, skin blemishes, blood splatters, high microbial contamination, spoilage, bone fractures, and mortality (Farouk et al., 2014). This highlights the vital need for appropriate animal management throughout the meat production process. Apart from the issue of unsanitary conditions in slaughterhouses, there is a concern regarding the sale of meat in unclean environments, including open shelters and unsanitary kiosk butcheries that fail to adhere to acceptable meat handling practices (Warriss, 1990).

Blockchain solutions can help us to ensure that stated and implied needs are met, by safety and quality assurance. Blockchain technology also enables the tracking of specific products in real-time, a capability that can contribute to the reduction of food waste (Patel et al., 2023). For instance, contaminated products can be swiftly and accurately traced, preventing them from reaching consumers, while safe foods can remain on the shelves, reducing unnecessary disposal into landfills. A study report (Tan et al., 2018) states that Walmart, a major player in the U.S. food market, responsible for 20% of all food sales, has successfully executed two blockchain pilot projects. In a pre-blockchain era, Walmart conducted a trace-back test on mangoes at one of its stores, which took six days, eighteen hours, and twenty-six minutes to trace the mangoes back to their source farm. With blockchain technology, Walmart can now provide consumers with all the desired information in a mere 2.2 seconds. In cases of disease outbreaks or contamination, this significant reduction in time from six days to seconds is invaluable. Food safety is verified through current and precise management of quality assurance certificates that can be securely collected and easily shared among involved stakeholders on the blockchain network (Tan et al., 2018). The immutability of blockchain data rationalizes audit procedures by minimizing the need for documentation, manual tasks, and the associated possibility of human error (Patel et al., 2023).

9.2.3 Certification and regulation

Quality accreditations are of utmost significance for retailers and consumers when assessing the excellence and safety of food. The conventional process of accreditation entails frequent and often time-consuming audits, which can prove to be burdensome when dealing with conventional paper-based documentation (Li et al., 2020). However, the advent of blockchain technology holds immense potential in revolutionizing the storage, management, and sharing of certifications with authorized parties (Antonucci et al., 2019). This is specifically relevant to the realm of organic and certified origin products. By implementing blockchain, a centralized and real-time record of accredited entities can be established, encompassing suppliers, farmers, processors, and certifying bodies (Cao et al., 2022). This process will facilitate the accreditation body in streamlining the issuance and management of accreditations. Moreover, certificate issuance will be conducted in a secure manner, while granting tailored viewing permissions to each participant in the network (Li et al., 2020). Thus, the introduction of blockchain will alleviate the complexity associated with auditing and approving certificates, thereby fostering trust among stakeholders.

The Food Standards Agency (FSA) in the UK exhibited the potential of blockchain technology. The FSA has successfully implemented two blockchain-based pilot projects in slaughterhouses as a regulatory tool within the food sector (Aronzon, 2020). Remarkably, the FSA's pilot initiatives mark the pioneering use of blockchain technology to verify regulatory compliance in the food sector. The pilot programs involved the use of a permissioned blockchain, which ensured that data could be accessed by both the FSA and the participating slaughterhouses. This approach significantly enhanced transparency, leading the FSA to express its intention to replicate the trial across the entire food supply chain.

9.2.4 Smart contracts

Smart contracts are programmable codes that execute the terms and conditions of a contract autonomously, relying on predetermined parameters. These smart contracts utilize blockchain technology, which has the potential to revolutionize the landscape of contract farming. The terms and conditions of contract farming are encoded within the blockchain, ensuring their irreversibility. By automating contracts, significant cost savings can be achieved for businesses through the reduction of transaction fees and legal expenses (Trivedi et al., 2023). These contractual agreements are executed, supervised, decentralized, and enforced autonomously within the blockchain server (Niranjanamurthy et al., 2019; Tredinnick, 2019). Real-time tracking of contractual obligations ensures prompt compliance with minimal delays, thereby reducing waiting times and minimizing uncertainties for all stakeholders (Bhat et al., 2021). These efficiency improvements have the potential to disrupt the traditional contract infrastructure and eliminate the need for centralized intermediaries.

Smart contracts enable direct communication between businesses, bypassing the requirement for intermediaries or third parties. These contracts securely store the agreed-upon terms and possess the capability to independently validate and enforce said terms, thereby enhancing the efficiency of the contractual process (Kumar et al., 2022). This shift in power dynamics not only empowers farmers but also reduces their reliance on middlemen, potentially leading to fairer prices for their produce. The monitoring of food will leverage a network of sensors that can track the entire journey of the produce from its origin to the consumer. Utilizing blockchain and smart contracts will ensure the integrity of data while also safeguarding the relationships of all parties involved in the supply chain (Tan et al., 2018). Smart contracts enable automatic transactions based on predetermined events, thereby eliminating the need for human communication, interactions, and the associated potential for errors typically found in such transactions.

9.2.5 Finance of trade

Livestock and corresponding product markets at global level are renowned for their dynamic nature, with a substantial percentage remaining domestic in scope and heavily influenced by local demand factors (McDermott et al., 2010). Consequently, the volume of milk and meat production engaged in international trade usually drifts around 10 percent (Chatellier, 2021). Managing these supply chains requires skillfully navigating the complexities associated with regulating and handling highly perishable goods, all while prioritizing food safety concerns (Lezoche et al., 2020). The elaborate structure of these supply chains often results in longer lead times and increased transaction costs (Sarpong, 2014). However, the emergence of blockchain technology offers a potential solution by streamlining agricultural supply chains, enhancing food safety measures, promoting inclusive trade, mitigating risks in trade finance, expanding access to financial services in agriculture, generating comprehensive market data, and establishing clear legal frameworks for land ownership (Niranjanamurthy et al., 2019).

The trade finance industry has encountered difficulties in meeting the requirements of Micro, Small, and Medium Enterprises (MSMEs) and emerging economies (Kimani et al., 2020). This lack of financing can hinder economic growth, with estimates suggesting a global trade finance deficit of $1.5 trillion as of 2017 (Dahdal et al., 2020). Traditionally, established players have enjoyed preferential access to trade finance, leaving MSMEs at a disadvantage (Javaid et al., 2022). However, DLT provides a promising solution, with the potential to mitigate bank risks, promote inclusivity, and expand the range of services available to MSMEs (Kimani et al., 2020). Increasingly, blockchain-based applications in the trade finance sector are targeting MSMEs and regions with limited access to traditional financing (Javaid et al., 2022). In agricultural supply chains, financial transactions are of great importance as they involve payments between different stakeholders such as farmers, processors, traders, exporters, and loan providers (Lezoche et al., 2020). However, cash-based transactions are the current norm, leading to inefficiencies, high costs, and risks like theft and loss (Javaid et al., 2022). Nevertheless, blockchain solutions offer a promising avenue for streamlined and instantaneous payment services.

9.2.6 Unified connectivity

Blockchain technology can seamlessly integrate with various business processes, simplifying them. Its incorporation into business infrastructure enhances accuracy, offers improved procurement visibility, and provides dependable data for analytics (Tran et al., 2021). Blockchain serves as an integrative technology, offering enhanced technological solutions to businesses and complementing Enterprise Resource Planning (ERP) software (Zafar et al., 2022). This integration maintains the same visibility of business processes and user interface but significantly widens the perspective on inventory (Vivaldini & De Sousa, 2021). Moreover, the technology can integrate all stakeholders under a single platform and fulfill the information requirement so as to reduce the bullwhip effect in the supply chain process (Tran et al., 2021). The blockchain technology has shown a wide adoption rate as far as the poultry sector is concerned (Alshehri, 2023). As the sector is well organized, it will be easier for the players to integrate the blockchain into the existing system. Carrefour SA, the largest retailer in Europe, has embraced blockchain ledger technology to monitor and trace the journey of chickens, eggs, and tomatoes as they make their way from farms to retail stores (Chang et al., 2020). They had implemented the system in collaboration with blockchain technology developed by IBM (IBM.N). IBM's blockchain technology solution enables the industry to track and share information about the cultivation, processing, and shipping of products (Aronzon, 2020). This technology dramatically reduces the time required to verify the origin of food products from days or weeks to mere seconds. Along similar lines, in India, Nandu's Chicken, which is subsidiary of Bengaluru-based Nanda Feeds Ltd, exploits a combination of Internet of Things (IoT), analytics, and blockchain, along with high-end equipment, to warrant that chicken meat matches the highest quality standards right from its initial stage as an egg.

Overall, the utilization of blockchain technology presents notable benefits to various industries, including the poultry sector, by improving transparency, precision, and accountability across the supply chain. By incorporating this technology into current systems, businesses can potentially transform their operations and gain valuable data for analysis and decision-making commitments.

9.2.7 Customer feedback

Over the past several years, there have been numerous food safety and animal disease crises across the globe, eroding consumer confidence in the safety of meat products and the regulatory authorities tasked with ensuring food security (Warriss, 1990). Blockchain, with its inherent feature of creating business transparency, has the potential to rebuild trust among consumers, as it relies on the visibility of data transfers. The Quick Response (QR) codes on product packaging grant consumers access to blockchain-stored information through augmented reality (Zafar et al., 2022). By scanning a QR code with their smart mobile device, consumers can access detailed product journey data. In response to the growth of online retail, hyperlinks to webpages can be added, allowing consumers to access blockchain information for informed purchase decisions (Patel et al., 2023). A blockchain-enabled feedback loop provides consumers with an effortless way to submit product reviews. This ensures that all participants in the food chain have access to authentic and unaltered consumer feedback, crucial for brands to stay competitive in highly competitive markets.

9.3 HOW BLOCKCHAIN TECHNOLOGY WORKS IN LIVESTOCK

As discussed earlier, the major problem the livestock industry facing is the traditional supply chain which causes the bullwhip effect as each link in the supply chain will over or underestimate the product demand resulting in exaggerated fluctuations and information distortion (Alshehri, 2023; Patel et al., 2023) As shown in Figure 9.1, in traditional supply chain, the animals/products are transferred from one party to another, each party also requires transferring financial and specification data records along the supply chain. The flow of information and the accuracy of the data transferred among the stakeholders depend upon the mutual trust between them.

Friction within supply chains is a significant issue, primarily due to an excessive number of intermediaries and the resulting back-and-forth complexities (Sarpong, 2014). This elevated level of uncertainty disrupts the efficient functioning of supply chains (Kumar et al., 2022). Instead of direct

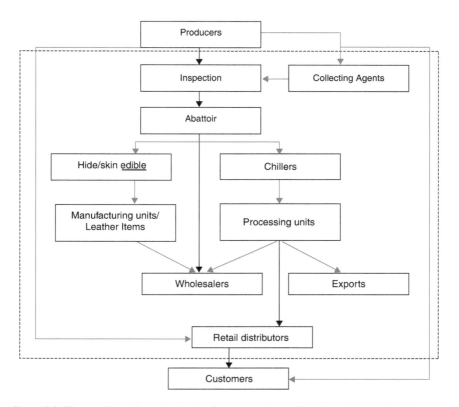

Figure 9.1 The traditional supply chain of meat products (Developed by authors)

interactions between suppliers, service providers, and clients, these parties often must engage with central third-party entities. Consequently, what should be straightforward transactions evolve into protracted, multi-step processes. Blockchain technology can be implemented to track physical products and authenticate their attributes from their origin to the point of sale (PoS). This application enhances supply chain transparency and traceability. Figure 9.2 illustrates a generic blockchain solution that has the potential to effectively maintain visibility throughout the livestock value chain (Patel et al., 2023).

The following steps describe how the blockchain system works in the livestock sector. Every animal in the farm is attached with a unique identifier such as a barcode, QR-code, or an RFID transmitter which enables the blockchain to track the particular animal (Kumar et al., 2022). Using their smartphones, farmers will input data about their farms and animals, interacting with blockchain technology in a straightforward manner, regardless of their familiarity with its intricacies. The smartphone interfaces are thoughtfully crafted for ease of use, catering to users with varying levels of technical knowledge. Generally, the farmer producer will feed the following fields of information to the blockchain:

   a. Farmer's/farm registered name and details
   b. Sex
   c. Type of animal/breed
   d. Animal's birth/arrival date
   e. Geo-tagged image of the farm
   f. Animal's health and dietary plan
   g. Base weight
   h. Farmer's quality assurance certificates

Once the farmer makes the decision to sell their livestock, they transfer ownership to the buying agent. The buying agent assumes the responsibility of inspecting and documenting the animals, recording essential details such as:

   a. Details of the farmer
   b. Other details of the animal
   c. Quantity procured along with weight of animal
   d. Date of procurement

The buying agent collates the animals from different farms in similar way and creates a consignment. The consignment then passes to the Abattoir. Once the consignment arrives at the abattoir, a veterinarian verifies the information, and the animals are moved to the slaughterhouse. After the inspection and routine checks, the animal is slaughtered as per the standard

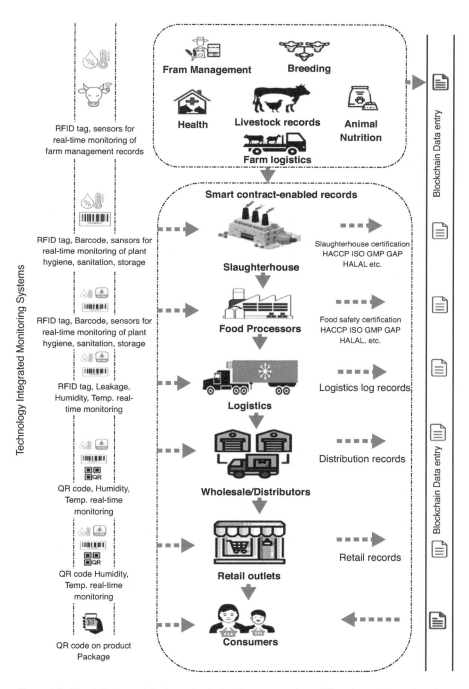

Figure 9.2 Blockchain-enabled supply chain of meat products [Developed by authors]

fixed criteria. Once the meat is prepared, a QR code is automatically generated. The QR code is embedded on the product wrapper with the following information:

a. Animal arrival date
b. Buying agents details
c. Farmer's name
d. Animal's health status
e. Expiry date of meat
f. Date of slaughter
g. Cuts made

The consignment is created and despatched to the wholesale/retailer or to the primary/secondary processors. Every consignment box is equipped with a magnetic reed switch that triggers an alarm if tampered with. The consignment box is additionally secured using a thermocouple, a polymeric sensor, and a passive RFID tag (Patel et al., 2023). Prior to shipment, the consignment's weight is verified and recorded in the blockchain. Each consignment is assigned a unique QR code, firmly associated with the QR codes of the products contained within it. This process initiates a smart contract for the consignment.

Throughout the transportation process, sensors continuously monitor the storage conditions of the consignment. The passive RFID tag records a time stamp at the commencement of the journey and another at its conclusion, typically at the cold storage warehouse or processing plant. Any deviations from the expected journey duration or storage conditions, as well as any signs of tampering, are treated as violations of the smart contract stored on the blockchain (Patel et al., 2023). For instance, if the consignment contains frozen meat with an expected temperature range of 28°F to 32°F throughout the journey, the smart contract will register a violation if the temperature strays from the specified standard at any point during transit. Similar parameters are also documented at the cold storage warehouse prior to the consignment's dispatch to the processing plant. Upon arrival at the processing plant, the consignment's weight is again assessed, and the results are recorded in the blockchain. This weight is then verified against the initial measurement, and a smart contract is executed accordingly. If any deviations from the specified standards are detected, the consignment is rejected. Upon arrival at the primary/secondary processing plant, employees have the capability to scan the QR code on each product package. This allows for a precise traceability of the produce back to its geo-tagged farm of origin.

Similarly, the end products feature QR codes that provide consumers with access to blockchain-stored information through an augmented reality interface. Using their smartphones, consumers can scan the QR code to access comprehensive details about the product's journey to the store shelf. Furthermore, comfortable, effortless, and incentivized feedback loop is also provided, enabling consumers to provide product reviews. After the consumer submits feedback, it will be posted on the blockchain and made accessible to all network participants. It's imperative to notice that once feedback is submitted by the customer, no network member can alter it. The data stays immutable on the blockchain server.

9.4 THE TE-FOOD INITIATIVE: PIONEERING BLOCKCHAIN IN LIVESTOCK

TE-FOOD, a groundbreaking farm-to-table traceability platform, has arisen as a leading solution in growing economies like Vietnam since its inception in 2016. This pioneering technology methodically tracks agricultural products throughout their journey, from farms to wholesalers, slaughterhouses, and retailers. TE-FOOD ensures the safety and quality of food by promoting transparency and accountability in the entire supply chain. This not only empowers businesses, customers, and regulatory bodies but also offers valuable insights into the origin and history of food (TE-FOOD, 2018). With an impressive customer base of 6,000 satisfied individuals, TE-FOOD effectively monitors the daily volume of various products, including 2.5 million eggs, 200,000 chickens, and 12,000 pigs. In response to the global concern of foodborne diseases, TE-FOOD proactively expanded its traceability capabilities in 2018 to encompass fish, cattle, fruits, seafood, and vegetables. This proactive action reinforces its dedication to safeguarding public health and fostering trust in the food system (TE-FOOD, 2018).

TE-FOOD employs a unique blockchain model that is openly accessible to the public but requires permission. This model allows both supply chain participants and consumers to operate master nodes, thereby decentralizing traceability data and providing consumers with exceptional visibility into the inner workings of the food system. The Token Forwarding Device (TFD) token, which is based on the Ethereum ERC20 standard, plays a vital role in promoting transparency (Motta et al., 2020). Its presence underscores TE-FOOD's commitment to empowering the food supply chain. TE-FOOD diligently tracks every stage of a product's journey from farms to vendors, equipping all stakeholders with valuable tools. Customers gain actionable insights into the origin and quality of their food, businesses navigate the supply chain with expert knowledge, and government organizations ensure adherence to regulations. This comprehensive approach yields numerous advantages, such as strengthened consumer trust, increased brand visibility, improved operational efficiency, prompt product recalls, seamless compliance with export requirements, and protection against counterfeit goods (TE-FOOD, 2018).

TE-FOOD's end-to-end system incorporates several essential elements. Identification tools hold 1D/2D barcodes, security seals, RFID ear tags, and label stickers (Motta et al., 2020). Traceability tools consist of B2B traceability administration mobile and web apps, a central system, external interfaces, and reporting tools (Motta et al., 2020). Retail and consumer tools include B2C fresh produce history insight mobile and web apps, along with retail-side food history digital signage tools. Furthermore, TE-FOOD offers national livestock management solutions, covering livestock administration and enforcement systems. The system also consists of several farm management tools, tailored to specific categories such as feeding, vaccination, and fertilizers. Blockchain incorporated supply chain map utilized by TE-FOODS is shown in Figure 9.3. Moreover, TE-FOOD integrates food safety tools, involving a fraud management system and a meat quality visual analysis system as well as food condition sensor equipment.

TE-FOOD's varied toolbox caters to various tracking needs. From tamper-proof QR labels to RFID tags, it amends your physical objects (TE-FOOD, 2018). This flexibility extends to how the data is accessed, with devoted apps for each interested party. Scanning of products happens on the go with the mobile app seamlessly integrates sensor data through the IoT Application Programming Interface (API) or delves deeper with the web tool. No change in software is needed – the Open Interface accommodates existing systems with flexibility (Motta et al., 2020). This adaptable style enhances accessibility and transparency all across the supply chain.

TE-FOOD offers application models that cater to a wide range of scales. The Private model allows individual organizations to meticulously monitor their internal operations with impressive precision (TE-FOOD, 2018). Conversely, the Institutional model permits public authorities and

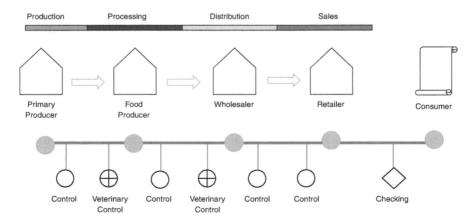

Figure 9.3 Blockchain technology integration in TE-FOOD Supply chain [Developed by authors]

regulators to gain a comprehensive understanding of entire industry sectors or geographical clusters (Motta et al., 2020). TE-FOOD's extensive range of offerings includes various applications and a complete set of tools that have been carefully designed to facilitate seamless traceability of food items at every stage of the supply chain. This solution guarantees end-to-end operational visibility and precise control over processes, serving the needs of both unique entities seeking internal traceability solutions and large, regulatory-driven applications that require compliance and oversight.

Despite facing challenges, TE-FOOD is actively navigating through them. Although the existing transaction speed of the TFD token requires optimization, ongoing development efforts are paving the way for future scalability (Motta et al., 2020). In a competitive environment, TE-FOOD continuously strives to innovate in order to maintain a leading position.

9.5 CONCLUSION

The livestock sector is presently in the early stages of implementing blockchain technology. However, industry managers are adamant to push the boundaries and investigate the full potential of blockchain-based solutions. Evidence shows that the blockchain technology is undergoing a rising adoption curve. This trend is not limited to developed nations; developing countries have also recognized its significant importance. Governments across several countries have realized both the necessity and advantages of these technologies. Consequently, their efforts to promote blockchain solutions are poised to fuel the market's growth.

9.5.1 Implications of the study

This research makes a valuable contribution to the existing body of knowledge by emphasizing the potential of blockchain technology in enhancing transparency and traceability within the agriculture and livestock sector. Through the exploration of specific use cases and the presentation of a case study, it advocates for practical benefits of integrating blockchain into the agri-food supply chain. As a result, this study expands our understanding of the applicability of blockchain and establishes it as a viable solution to address the limitations of the current supply chain system.

The implications of this study are significant for agri-food businesses and stakeholders. By adopting blockchain technology, companies could enhance transparency and traceability at all stages of the supply chain. This facilitates them to meet the growing demand for information regarding the source, build trust with consumers, and ensure quality of agricultural products. Furthermore, employing blockchain can help alleviate risks, improve operational efficiency reduce fraud. Therefore, managers should consider investing in blockchain-based solutions and foster collaboration with partners in order to establish comprehensive integration frameworks and standardized protocols.

9.5.2 Limitations and future scope

While this study attempts to delve into insights within the livestock sector, it acknowledges certain limitations that prompt further investigation. Primarily, the analysis predominantly focuses on TE-FOOD, which serves as an effective case study but possibly not representative of the varied challenges and occurrences across the entire agri-food system. Furthermore, the article recognizes the necessity to surpass the mere showcasing of the advantages of blockchain technology. A critical examination of imminent drawbacks of blockchain technology, including interoperability, regulatory obstacles, and scalability, would facilitate a more nuanced understanding of its real-world implementation.

This leads us to future research prospects in this domain. One key scope lies in investigating the applicability of blockchain in other sectors of the agricultural supply chain which have been relatively less explored, such as crop production, processing, and distribution. This endeavour will extend our comprehension of its capabilities across the entire food supply chain. Surmounting these obstacles will be vital in unleashing the full potential of blockchain technology in revolutionizing the agri-food landscape.

REFERENCES

Alshehri, Dr. M. (2023). Blockchain-assisted internet of things framework in smart livestock farming. *Internet of Things*, 22, 100739. https://doi.org/10.1016/j.iot.2023.100739.

Antonucci, F., Figorilli, S., Costa, C., Pallottino, F., Raso, L., & Menesatti, P. (2019). A review on blockchain applications in the agri-food sector. *Journal of the Science of Food and Agriculture*, 99(14), 6129–6138. https://doi.org/10.1002/jsfa.9912.

Aronzon, S. M. (2020). Blockchain and geographical indications: A natural fit? *Journal of Intellectual Property Law & Practice*, 15(11), 913–925. https://doi.org/10.1093/jiplp/jpaa151.

Bastian, J., & Zentes, J. (2013). Supply chain transparency as a key prerequisite for sustainable agri-food supply chain management. *The International Review of Retail, Distribution and Consumer Research*, 23(5), 553–570. https://doi.org/10.1080/09593969.2013.834836.

Bhat, S. A., Huang, N.-F., Sofi, I. B., & Sultan, M. (2021). Agriculture-food supply chain management based on blockchain and IoT: A narrative on enterprise blockchain interoperability. *Agriculture*, 12(1), 40. https://doi.org/10.3390/agriculture12010040.

Bocquier, F., & González-García, E. (2010). Sustainability of ruminant agriculture in the new context: Feeding strategies and features of animal adaptability into the necessary holistic approach. *Animal*, *4*(7), 1258–1273. https://doi.org/10.1017/S1751731110001023.

Bosona, T., & Gebresenbet, G. (2013). Food traceability as an integral part of logistics management in food and agricultural supply chain. *Food Control*, *33*(1), 32–48. https://doi.org/10.1016/j.foodcont.2013.02.004.

Cao, Y., Yi, C., Wan, G., Hu, H., Li, Q., & Wang, S. (2022). An analysis on the role of blockchain-based platforms in agricultural supply chains. *Transportation Research Part E: Logistics and Transportation Review*, *163*, 102731. https://doi.org/10.1016/j.tre.2022.102731.

Chang, Y., Iakovou, E., & Shi, W. (2020). Blockchain in global supply chains and cross border trade: A critical synthesis of the state-of-the-art, challenges and opportunities. *International Journal of Production Research*, *58*(7), 2082–2099. https://doi.org/10.1080/00207543.2019.1651946.

Chatellier, V. (2021). Review: International trade in animal products and the place of the European Union: main trends over the last 20 years. *Animal*, *15*, 100289. https://doi.org/10.1016/j.animal.2021.100289.

Dabbene, F., Gay, P., & Tortia, C. (2014). Traceability issues in food supply chain management: A review. *Biosystems Engineering*, *120*, 65–80. https://doi.org/10.1016/j.biosystemseng.2013.09.006.

Dahdal, A., Truby, J., & Botosh, H. (2020). Trade finance in Qatar: Blockchain and economic diversification. *Law and Financial Markets Review*, *14*(4), 223–236. https://doi.org/10.1080/17521440.2020.1833431.

Dandage, K., Badia-Melis, R., & Ruiz-García, L. (2017). Indian perspective in food traceability: A review. *Food Control*, *71*, 217–227. https://doi.org/10.1016/j.foodcont.2016.07.005.

Danezis, G. P., Tsagkaris, A. S., Brusic, V., & Georgiou, C. A. (2016). Food authentication: State of the art and prospects. *Current Opinion in Food Science*, *10*, 22–31. https://doi.org/10.1016/j.cofs.2016.07.003.

El Koshiry, A., Eliwa, E., Abd El-Hafeez, T., & Shams, M. Y. (2023). Unlocking the power of blockchain in education: An overview of innovations and outcomes. *Blockchain: Research and Applications*, *4*(4), 100165. https://doi.org/10.1016/j.bcra.2023.100165.

Farouk, M. M., Al-Mazeedi, H. M., Sabow, A. B., Bekhit, A. E. D., Adeyemi, K. D., Sazili, A. Q., & Ghani, A. (2014). Halal and kosher slaughter methods and meat quality: A review. *Meat Science*, *98*(3), 505–519. https://doi.org/10.1016/j.meatsci.2014.05.021.

Hauge, S. J., Nafstad, O., Røtterud, O.-J., & Nesbakken, T. (2012). The hygienic impact of categorisation of cattle by hide cleanliness in the abattoir. *Food Control*, *27*(1), 100–107. https://doi.org/10.1016/j.foodcont.2012.03.004.

Hobbs, J. E. (2004). Information asymmetry and the role of traceability systems. *Agribusiness*, *20*(4), 397–415. https://doi.org/10.1002/agr.20020.

Javaid, M., Haleem, A., Singh, R. P., Suman, R., & Khan, S. (2022). A review of Blockchain Technology applications for financial services. *BenchCouncil Transactions on Benchmarks, Standards and Evaluations*, *2*(3), 100073. https://doi.org/10.1016/j.tbench.2022.100073.

Jiang, G., Hu, H., & Wang, Y. (2023). How do livestock environmental regulations promote the vertical integration of the livestock industry chain? Evidence from Chinese-listed livestock enterprises. *Journal of Cleaner Production, 413*, 137508. https://doi.org/10.1016/j.jclepro.2023.137508.

Jiang, J., & Xiong, Y. L. (2015). Technologies and mechanisms for safety control of ready-to-eat muscle foods: An updated review. *Critical Reviews in Food Science and Nutrition, 55*(13), 1886–1901. https://doi.org/10.1080/10408398.2012.732624.

Kache, F., & Seuring, S. (2017). Challenges and opportunities of digital information at the intersection of Big Data Analytics and supply chain management. *International Journal of Operations & Production Management, 37*(1), 10–36. https://doi.org/10.1108/IJOPM-02-2015-0078.

Khan, W., Jamshed, M., & Fatima, S. (2020). Contribution of agriculture in economic growth: A case study of West Bengal (India). *Journal of Public Affairs, 20*(2), e2031. https://doi.org/10.1002/pa.2031.

Kimani, D., Adams, K., Attah-Boakye, R., Ullah, S., Frecknall-Hughes, J., & Kim, J. (2020). Blockchain, business and the fourth industrial revolution: Whence, whither, wherefore and how? *Technological Forecasting and Social Change, 161*, 120254. https://doi.org/10.1016/j.techfore.2020.120254.

Kumar, A., Srivastava, S. K., & Singh, S. (2022). How blockchain technology can be a sustainable infrastructure for the agrifood supply chain in developing countries. *Journal of Global Operations and Strategic Sourcing, 15*(3), 380–405. https://doi.org/10.1108/JGOSS-08-2021-0058.

Lezoche, M., Hernandez, J. E., Alemany Díaz, M. D. M. E., Panetto, H., & Kacprzyk, J. (2020). Agri-food 4.0: A survey of the supply chains and technologies for the future agriculture. *Computers in Industry, 117*, 103187. https://doi.org/10.1016/j.compind.2020.103187.

Li, J., Maiti, A., Springer, M., & Gray, T. (2020). Blockchain for supply chain quality management: Challenges and opportunities in context of open manufacturing and industrial internet of things. *International Journal of Computer Integrated Manufacturing, 33*(12), 1321–1355. https://doi.org/10.1080/0951192X.2020.1815853.

Liu, Y., Pan, X., & Li, J. (2015). Current agricultural practices threaten future global food production. *Journal of Agricultural and Environmental Ethics, 28*(2), 203–216. https://doi.org/10.1007/s10806-014-9527-6.

Loizou, E., Karelakis, C., Galanopoulos, K., & Mattas, K. (2019). The role of agriculture as a development tool for a regional economy. *Agricultural Systems, 173*, 482–490. https://doi.org/10.1016/j.agsy.2019.04.002.

Mao, D., Wang, F., Hao, Z., & Li, H. (2018). Credit evaluation system based on blockchain for multiple stakeholders in the food supply chain. *International Journal of Environmental Research and Public Health, 15*(8), 1627. https://doi.org/10.3390/ijerph15081627.

McDermott, J. J., Staal, S. J., Freeman, H. A., Herrero, M., & Van De Steeg, J. A. (2010). Sustaining intensification of smallholder livestock systems in the tropics. *Livestock Science, 130*(1–3), 95–109. https://doi.org/10.1016/j.livsci.2010.02.014.

Minoli, D., & Occhiogrosso, B. (2018). Blockchain mechanisms for IoT security. *Internet of Things, 1–2*, 1–13. https://doi.org/10.1016/j.iot.2018.05.002.

Motarjemi, Y., & Käferstein, F. (1999). Food safety, hazard analysis and critical control point and the increase in foodborne diseases: A paradox? *Food Control, 10*(4–5), 325–333. https://doi.org/10.1016/S0956-7135(99)00008-0.

Motta, G. A., Tekinerdogan, B., & Athanasiadis, I. N. (2020). Blockchain applications in the agri-food domain: The first wave. *Frontiers in Blockchain*, 3, 6. https://doi.org/10.3389/fbloc.2020.00006.

Nakamoto, S. (2008). Bitcoin: A peer-to-peer electronic cash system. https://www.ussc.gov/sites/default/files/pdf/training/annual-national-training-seminar/2018/Emerging_Tech_Bitcoin_Crypto.pdf.

Niranjanamurthy, M., Nithya, B. N., & Jagannatha, S. (2019). Analysis of blockchain technology: Pros, cons and SWOT. *Cluster Computing*, 22(S6), 14743–14757. https://doi.org/10.1007/s10586-018-2387-5.

Patel, A. S., Brahmbhatt, M. N., Bariya, A. R., Nayak, J. B., & Singh, V. K. (2023). Blockchain technology in food safety and traceability concern to livestock products. *Heliyon*, 9(6), e16526. https://doi.org/10.1016/j.heliyon.2023.e16526.

Sarpong, S. (2014). Traceability and supply chain complexity: Confronting the issues and concerns. *European Business Review*, 26(3), 271–284. https://doi.org/10.1108/EBR-09-2013-0113.

Sheng, L., & Zhu, M. (2021). Practical in-storage interventions to control foodborne pathogens on fresh produce. *Comprehensive Reviews in Food Science and Food Safety*, 20(5), 4584–4611. https://doi.org/10.1111/1541-4337.12786.

Smith, G. C., Tatum, J. D., Belk, K. E., Scanga, J. A., Grandin, T., & Sofos, J. N. (2005). Traceability from a US perspective. *Meat Science*, 71(1), 174–193. https://doi.org/10.1016/j.meatsci.2005.04.002.

Tan, B., Yan, J., Chen, S., & Liu, X. (2018). The impact of blockchain on food supply chain: The case of Walmart. In M. Qiu (Ed.), *Smart Blockchain* (Vol. 11373, pp. 167–177). Springer International Publishing. https://doi.org/10.1007/978-3-030-05764-0_18.

TE-FOOD. (2018). Making business profit by solving social problems. https://te-food.com/wp-content/uploads/2020/11/te-food-white-paper.pdf.

Tran, N. K., Ali Babar, M., & Boan, J. (2021). Integrating blockchain and internet of things systems: A systematic review on objectives and designs. *Journal of Network and Computer Applications*, 173, 102844. https://doi.org/10.1016/j.jnca.2020.102844.

Tredinnick, L. (2019). Cryptocurrencies and the blockchain. *Business Information Review*, 36(1), 39–44. https://doi.org/10.1177/0266382119836314.

Trivedi, C., Rao, U. P., Parmar, K., Bhattacharya, P., Tanwar, S., & Sharma, R. (2023). A transformative shift toward blockchain-based IoT environments: Consensus, smart contracts, and future directions. *Security and Privacy*, 6(5), e308. https://doi.org/10.1002/spy2.308.

Vivaldini, M., & De Sousa, P. R. (2021). Blockchain connectivity inhibitors: Weaknesses affecting supply chain interaction and resilience. *Benchmarking: An International Journal*, 28(10), 3102–3136. https://doi.org/10.1108/BIJ-10-2020-0510.

Warriss, P. D. (1990). The handling of cattle pre-slaughter and its effects on carcass and meat quality. *Applied Animal Behaviour Science*, 28(1–2), 171–186. https://doi.org/10.1016/0168-1591(90)90052-F.

Zafar, S., Bhatti, K. M., Shabbir, M., Hashmat, F., & Akbar, A. H. (2022). Integration of blockchain and internet of things: Challenges and solutions. *Annals of Telecommunications*, 77(1–2), 13–32. https://doi.org/10.1007/s12243-021-00858-8.

Chapter 10

The role of blockchain in the sustainability of green technology

Rageshree Sinha, Sushil Mohan, and Arvind Upadhyay

10.1 INTRODUCTION

Blockchain technology is a revolutionary decentralised system that has redefined how we handle digital transactions and data. At its core, blockchain is a distributed ledger that operates on a network of computers, allowing for the secure and transparent recording of transactions. Blockchain has some evolved features as per the details in the next section, which makes it a suitable technology for enhancing the green and sustainability aspects of businesses. There have been many studies that explore the features of blockchain technology in different sectors and how it's used to make business models greener and more sustainable. This study captures the very important research gap, which leads to comparing the features of the technology across multiple sectors to evaluate how useful it is for making a green and sustainable economy. Blockchain is not limited to cryptocurrencies; its applications extend across industries, offering solutions for supply chain management, smart contracts, decentralised finance, and much more. As a transformative technology, blockchain has the potential to redefine the way we interact with data, enabling greater transparency, efficiency, and trust in various aspects of our digital world. The literature review studies the evolved features of this technology. Then the literature review compares the sectors and discusses the potential for the technology to make a green and sustainable economy. This chapter delves into the practical aspects of implementing and adopting blockchain, for understanding how its evolved features are being utilised in various sectors. These are some of the major focus areas today from the point of view of climate emergency.

Hence in this study, the first part will connect the concept of blockchain technology as a green and sustainable technology concerning electricity usage and consumption in the domestic market. The second part is a literature review which will establish the areas of efficiency observed and connect it to the current research gap. This is followed by the data analysis and the discussion section. The study is summarised into a conceptual framework which is concluded in the conclusion section.

10.2 LITERATURE REVIEW

The blockchain technology has immense capability to be utilised in different areas, other than just electricity for domestic use. As every nation approaches the net zero targets, new and emerging technologies and their evolving features must be utilised to achieve our targets. Blockchain certainly is the focus now with its transformative capabilities. In this literature review section, the main focus will be the literature that lays emphasis on blockchain – being the green technology and its various applications to ensure the reduction of the carbon footprint to make it more sustainable.

10.2.1 The transforming features of blockchain technology

The concept of Blockchain is a disruptive computing paradigm, and it follows the computer mainframes, Internet, social networking platforms, and the mobile (Swan, 2015). From climate perspectives for the energy sector, some of the features that help are exploring alternatives to energy-intensive proof-of-work (PoW) algorithms, such as proof-of-stake (PoS) or delegated proof-of-stake (DPoS).

The feature of tokenisation is observed to be used to design carbon trading. This is through the introduction of the carbon coin. This is considered a blockchain asset as it provides the right of energy producers to emit carbon (Golding et al., 2022). The socio-economic trend observed now is that every consumer desires to use a product that has less carbon footprint. Blockchain is capable of being used for sharing this information with the end users. This becomes possible because of its distributed technology (da Cruz et al., 2020). The study conducted by (da Cruz et al., 2020) uses the smart contract feature to create track and traceability of the carbon footprints to its producers. This is a very critical and essential requirement as the world is joining hands to control the increase in the temperature to 1.5 degrees centigrade. Being of the distributed nature, the distributed application (DApp) platform has been initiated to support the layers of architecture. The smart contract is supported by the Ethereum blockchain. The Node.js also supports the smart contract (da Cruz et al., 2020; Wood, 2014). Concerning smart contracts for electricity usage, the observed phenomenon is that market users have increased the variety of transactions, and this is the base of flexible trading contracts (Lu et al., 2021. The other aspects of blockchain technology that are very important to consider are decentralisation, transparency, and autonomy combined with anonymity (Lin & Liao, 2017). The bigger picture shows the aim to establish a decentralised automation energy (DAE) community (Pop et al., 2018). Managing the demand response programs by using the decentralised feature of blockchain based on the peer-to-peer systems facilitates the seamless enhanced operation processes in the current grid (Hwang et al., 2017).

The fashion industry and carbon footprint connect when fast fashion is considered. Blockchain technology has been evaluated to create more environmentally sustainable solutions in the apparel industry (Fu et al., 2018). The emission trading system (ETS) is the novel emission link which is helpful to convey the carbon emission to the public for each of the key steps for the clothing manufacturing process. This helps in establishing features to reduce the emission in each step (Fu et al., 2018). In other words, implementing an ETS and linking it with blockchain technology can bring transparency, traceability, and efficiency to the carbon market. The further research path for this study is to apply the analytical skills to derive the energy wastes and carbon emission evaluation. This includes a lot of other skill sets which are more blockchain technology-based skill development. While this technology gets implemented to create a more sustainable outcome, the integrity of the information is taken into crucial consideration and is evaluated to be protected (Jakobsson & Juels, 1999). The steps included in the operation are transparent and are immune to contamination. Various other features are the distribution, transfer, cancellation, and inventory that manage the carbon allowance is managed efficiently.

The Ethereum technology features enhance the blockchain operation platforms. This has resulted in a huge surge of blockchain-based vehicle charging stations (Dodge, 2015; Wood, 2014). Carbon emission has been a major consideration in the manufacturing process post the Kyoto Protocol (Ma et al., 2023). It's worth mentioning that carbon emission trading has been like commodity trading, and it was visibly more potential and profitable. The socio-economic impacts that influence this (Khaqqi et al., 2018; Lin et al., 2017). According to Ma et al., (2023), blockchain technology can be considered for being a green technology as it can be blended with new and emerging technologies like Internet of Things (IoT) to create a reward system that leads to efficient energy usage. This can be an important landmark for the energy sector that in turn reduces carbon emissions.

There is a significant carbon emission of greenhouse gasses from the building sector (Can & Price, 2008). Some of the factors that contribute to an exponential increase, an estimated 50% increase in this carbon emission are urbanisation, pollution, and climate emergency factors (EIA & Kahan, 2019; Moazami et al., 2019). The trend and pattern observed in the usage of energy and the carbon-di-oxide emission (du-can & Price, 2008). This building sector does observe a wide usage behaviour and large variance of the electrical energy consumption (Hong et al., 2017; Ma et al., 2023; Sikorski et al., 2017). The peak time of energy consumption was observed and some of the factors that came to light are the general demographics like age of the population and other variables of the gross domestic product (du-can & Price, 2008). According to Ma et al., (2023), blockchain technology has been used to calculate and manage building sector energy consumption to present an innovative and transparent approach. Implementing smart metering systems in conjunction with the IoT devices to collect real-time data on energy consumption within buildings and utilising the blockchain to securely record and timestamp energy data from smart meters, smart grids, ensuring transparency, and preventing tampering to tokenise for establishing a reward system. The word tokenisation means tokenise energy consumption data, creating digital tokens that represent specific units of energy consumed by a building. These tokens can be traded on a blockchain-based marketplace, allowing for transparent and traceable transactions related to energy consumption. The other features pertaining to blockchain are smart contracts and automated transactions. The other observable features for the building sectors with respect to energy consumption are the use of smart contracts to enforce energy agreements, smart grids that automatically execute transactions based on predefined conditions, and facilitate transparent billing processes. Decentralisation observed in the energy markets is driven by the specific domain knowledge blended with practical reasons (Halu et al., 2016; Khalipour & Vassallo, 2015; Mylrea and Gourisetti, 2017). Moreover, the implementation of the microgrids involves the various wide range of medium- and low-voltage. The mention of spatial heterogeneities in urban areas creates fluctuations in the microgrids (Barthélemy, 2011). The decentralised control strategies that react to the system conditions can create more sustainable impacts (Shah et al., 2011). The value-adding service is the smart contracts in this energy and electricity sector. The reduction in the trust cost in the electricity can happen when the evidence of running the smart contract insurance in the peer-to-peer network which consists of the 4000 nodes shows the success rate of least transaction time to 16 secs. This not only reduces the trust cost but also increases the efficiency of the electricity transaction and charge settlement (Lu et al., 2021; Sikorski et al., 2017)..

10.2.2 Knowledge management and the transforming capabilities of blockchain technology

The implementation of blockchain technology is based on the knowledge that is shared and gained while designing the implementation. Some of the knowledge included are data acquisition, data processing, data management, and data provision (Hong et al., 2017; Worner et al., 2019). The knowledge manifestation regarding blockchain technology implementation will be critical for a high-performing building. Analytical techniques like the knowledge of big data and data mining will create efficiency over the shortcomings of the traditional techniques (Hong et al., 2017). It has been emphasised that detailed knowledge of the actual distribution can help make informed decisions (Halu et al., 2016). Sharing of information to provide the knowledge base as a bridge between the companies and consumers is essential to establish (da Cruz et al., 2020). According to Jakobsson and

Jules (1999), knowledge is essential to establish the Proof-of-Work (PoW) in the cryptographic protocols. Leveraging learning techniques will be useful in generating responses to signals from the network. This will be useful for monitoring the energy consumption. It's important to note that the successful implementation of blockchain in knowledge management requires careful consideration of specific used cases, organisational needs, and user engagement strategies.

From the above literature, it's evident that Blockchain technology is using its enhanced features to create more eco-friendly methods in every sector for more sustainable platforms. The study of this literature body creates the pathway to understanding the role of blockchain technology implementation in creating more green and sustainable platforms which can be integrated with other new and emerging technologies.

This body of literature essentially establishes the areas in which the transforming capabilities of blockchain technology are being used to make more green outcomes and sustainable sectors in the long run.

10.3 DATA ANALYSIS AND DISCUSSION

From the above literature collected across different sectors, the features that are enabling blockchain technology to become greener and create sustainable solutions are mapped. This section is divided into two main sections. The first part shortlists the method of selecting the articles and the second section selects the data that will help to understand and answer the research question that it is a technology which is being implemented in most of the sector.

10.3.1 PRISMA flow chart of selecting articles

The method of selecting the articles is called the PRISMA flow chart (Moher et al., 2009). This process helps in creating the detailed selection process structure. The keyword search has shown a total of 5286 articles. There are 160 articles from the Science Direct, 220 from the Elsevier, 4893 from the Google Scholar, 4 from Wiley, and 9 from the ResearchGate. A total of 5286 articles have been then narrowed down by eliminating any articles before 1990. After eliminating 1978 articles, there remains 3308 articles from the identification stage. The next screening phase removes articles that are non-researched – 1410 articles, non-peer-reviewed articles – 324 articles, and non-business management and social science articles – 326 articles. The post-screening stage leaves behind the 1248 articles. The eligibility is done after a full review of the articles. This stage eliminates articles which are not based on the blockchain features – 1012 articles. It also eliminates articles that are not relevant about how the blockchain is the Green Technology for sustainability – 207 articles. The 29 relevant articles which are finally left are included as they are relevant articles that establish the features that make blockchain different. It speaks about the role of the various features that can be utilised in various industries. This helps target the established research aim of understanding the underlying role of knowledge to enhance the implementation and reap the competitive advantage. The competitive advantage will be achieved in various ways as per the customer needs but in addition, it creates a pathway for achieving the net zero goals. This is shown in Figure 10.1:

10.3.2 Blockchain implementation as per the sectors

The next part of the section investigates the literature that discusses the blockchain feature as utilised in the sectors. This provides an insight into the most popular sector utilising the special blockchain features to be strategically more sustainable. The energy sector shows that implementation of the blockchain is highest in the energy sector and collaboration works with the building sector. This is followed by the IT and security sector followed by the articles that involve the economy. Technology, agriculture, fast fashion, and manufacturing have a similar number of articles each. This means that every sector has used blockchain as a potential green technology. Efforts have been observed to understand that some features of blockchain technology can reduce carbon footprint, some features can measure carbon emission, and some can monitor the consumer behaviour pattern to align with the net zero goals and objectives. The technology with the evolved features does have the potency to counter the climate emergency that we are facing today. When utility and energy sectors become green and more granularised to customer needs, they become more sustainable too. The long run impacts for this sector. The fast fashion sector, manufacturing sector, agriculture IT, and security sectors are also seeing the implementation of the blockchain in various areas. The fast fashion sector which is very much responsible for enhancing the impacts of climate change rapidly can now be monitored by blockchain technology. This will be effective in creating less carbon emission and footprint (Figure 10.2).

10.3.3 Features of blockchain technology

Below is a representation of blockchain features in a word cloud. The visualisation has been created with the help of the Power BI. The most popular feature that enables blockchain to become a green technology is the smart contracts. The tokenisation follows this. The features highlighted in the word cloud are responsible for creating a more sustainable environment by reducing the carbon footprint. These features help in customising the needs of the customer and hence minimise the carbon footprint in the long run (Figure 10.3).

204 Blockchain Technology

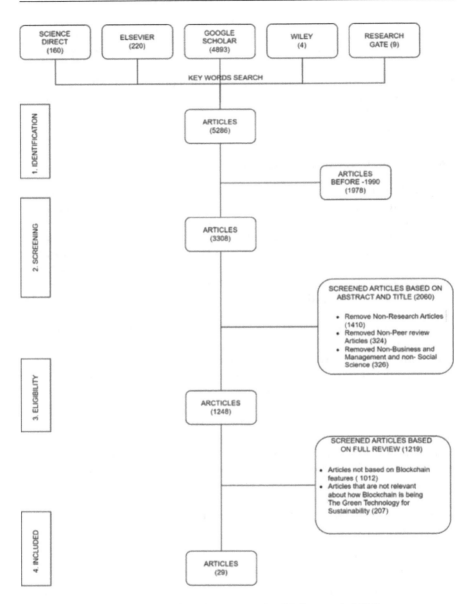

Figure 10.1 PRISMA flow chart for selecting articles (Moher et al., 2009).

The below graph shows the number of articles that are published with respect to blockchain technology and its evolved features. The highest number of articles was observed in 2017. As observed from the patterns, this is the time when blockchain technology has been implemented. Most of the articles that provide evidence of sustainability are very recent. The last

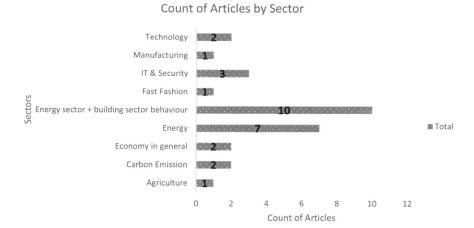

Figure 10.2 Blockchain technology implemented as per sector.

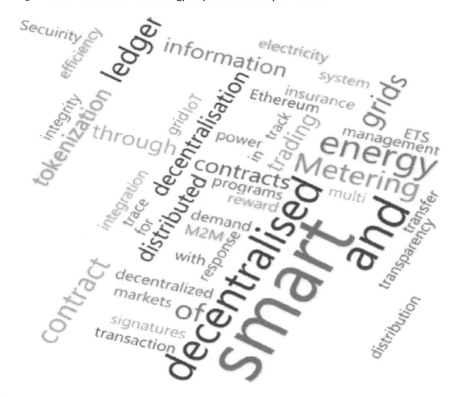

Figure 10.3 Word cloud with blockchain features that make the technology green and sustainable.

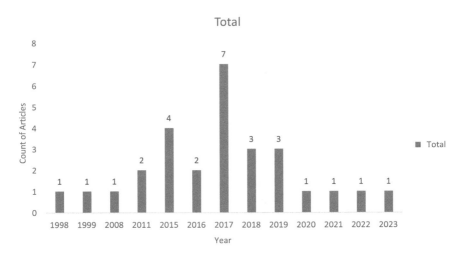

Figure 10.4 Articles as per year of publishing.

decade has seen the climate emergency as a high priority and the articles do establish that there have been lots of thoughts that have gone into using blockchain features to counter the carbon emissions and the carbon footprint. The last decade has seen the highest number of articles (Figure 10.4).

10.3.4 Discussion

The data above clearly shows that Blockchain has been implemented rapidly in the last few years, especially for its evolved features. The implementation is observed in most of the sectors. In this study, only a few sectors that are utilising the evolved features of blockchain technology are being considered. Among the collected articles, the energy sector in collaboration with the building sector is observed to be reaping the benefits of the technology. The net zero being the most influential socio-economic trend, blockchain technology features generate a competitive advantage in this sector. Blockchain technology has the potential to revolutionise the energy sector by addressing various challenges related to transparency, efficiency, and decentralisation. Through decentralised energy grids, blockchain can be implemented to enable peer-to-peer energy trading. Utilise smart contracts to automate and enforce energy trading agreements, ensuring secure and transparent transactions between parties. The theoretical and academic interest in these features creates a comparative analysis of using the technology and rendering the sector greener and more sustainable. This can improve grid stability by optimising energy distribution, reducing losses, and enhancing overall efficiency. The real-time data sharing can be employed by blockchain to facilitate real-time data sharing among different components of the energy grid, allowing for quick responses to fluctuations in supply and demand. Moreover, tokenise renewable energy generation and attribute certificates on the blockchain. This allows for transparent tracking of renewable energy production and simplifies the trading of Renewable Energy Certificates (RECs). Leverage the immutability of blockchain to create tamperproof records of renewable energy generation, providing credibility to REC transactions. The trace and tracking implement blockchain to trace the origin and production methods of energy. Consumers can verify the source of their energy, ensuring it comes from renewable and sustainable sources. The evolved features have immense practical implications as they lead to saving costs, promote efficiency, and in turn help the cause of climate emergency in the process. This is particularly useful in remote areas where traditional grid infrastructure is limited. The use of blockchain enables smart grids, smart metering and automated payments in off-grid or microgrid scenarios, reducing the need for centralised billing systems (Energy Blockchain Labs, 2018, Myrea and Gourisetti, 2017). Blockchain's ability to provide transparency, security, and decentralisation makes it a powerful tool for reshaping the energy sector and plays an important role in addressing climate emergencies. As the industry continues to transition towards cleaner and more sustainable practices, blockchain technology can play a pivotal role in driving innovation and efficiency.

The practical impact in the fashion industry is the implementation of blockchain technology to track the carbon footprint of fashion products. This includes recording the environmental impact of manufacturing processes, transportation, and materials used. Tokenisation is also observed in the fashion industry. Consumers can use these tokens to make informed choices about supporting eco-friendly brands. Mostly tokenisation is the reward-based pathway. Implementing blockchain-based loyalty programs where consumers earn tokens for sustainable choices, such as recycling or purchasing eco-friendly products. Smart contracts are also observed in this industry. Implementing smart contracts to automate royalty payments to designers whenever their designs are sold or used. The practical implications of implementing blockchain are to streamline supply chain financing by creating transparent and traceable transactions. By integrating blockchain technology in these ways, the fashion industry can enhance transparency, sustainability, and trust, thereby addressing some of the critical challenges it faces. Blockchain has the potential to create a more ethical, transparent, and efficient fashion ecosystem, benefiting both consumers and industry stakeholders. Mostly these are the similar features used in the manufacturing and agriculture and the IT and security sectors as well.

10.4 CONCEPTUAL FRAMEWORK

From the features investigated above, the conceptual framework that can be formed from the above literature is as per the diagram below. There is an element of knowledge sharing that has been mentioned which facilitates

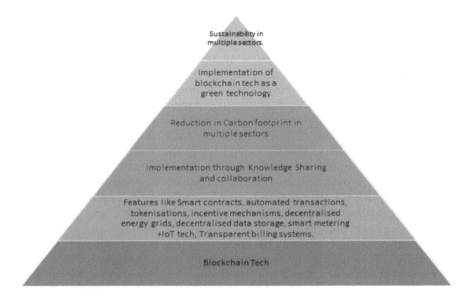

Figure 10.5 Conceptual framework of the Blockchain technology as a green technology.

blockchain technology to become a green technology and support sustainability across various sectors (Figure 10.5).

In other words, the layout above shows in a nutshell that blockchain can lead to sustainability. The features of blockchain technology like Smart metering and IoT integration, Decentralised data storage and validation, Smart contracts and automated transactions, Tokenisation of energy consumption, Transparent billing and payment systems, Decentralised energy grids, and Incentive mechanisms for energy efficiency can transform the capabilities in every sector (Fu et al., 2018; Ma et al., 2023; Moazami et al., 2019). The capabilities to implement processes that can lead to a reduction in the carbon footprint. When a technology adopts transforming capabilities, the impacts are visible in some very critical areas. For example, if decentralisation and trust are considered the impact observed is eliminating the need for a central authority. These fosters trust among participants as transactions are recorded on a distributed ledger that all parties can access and verify (Aune et al., 2018). The transforming capability here is the secure and transparent peer-to-peer transactions without relying on intermediaries. When the immutable and tamperproof records features are considered, data once recorded on the blockchain cannot be altered, providing immutability and a tamperproof record of transactions (Jakobsson & Juels, 1999; Lin & Liao, 2017). This capability ensures data integrity, transparency, and a reliable audit trail for regulatory compliance. When transparency and visibility are considered, the impact observed is regarding all participants to view and verify transactions, enhancing visibility across the supply chain or business processes. The capability transforming here facilitates real-time monitoring, traceability, and accountability. The smart contracts automation can impact the execution of predefined rules when specified conditions are met, reducing the need for intermediaries (da Cruz et al., 2020). The transforming capabilities observed here are streamlining and automating complex business processes, reducing costs and improving efficiency. The impact of tokenisation or digital tokens on the blockchain represents ownership or rights to real-world assets, facilitating fractional ownership and enhancing liquidity. The transforming capabilities here are the new possibilities for trading, investment, and financing in a wide range of assets. The impact of the enhanced security is observed in the cryptographic techniques to secure transactions, protecting against fraud and unauthorised access. The transforming capabilities observed are the robust security features, making blockchain networks resistant to hacking and ensuring data confidentiality. The impact of blockchain interoperability allows different blockchain networks to communicate seamlessly, fostering collaboration (Rauter, 2018). The transforming capability observed here is the connectivity between different systems and networks, facilitating data exchange that leads the interoperability. One significant impact observed in implementing the blockchain technology is the eradication of the middlemen. This can be one important capability that can provide the firm with direct peer-to-peer interactions become feasible, particularly in sectors such as finance, real estate, and supply chain. The impact of improved identity management features in blockchain technology provides a secure and decentralised way to manage digital identities (Aitzhan & Svetinovic, 2016). The capabilities transformed in this process have control over their personal data, reducing the risk of identity theft and enhancing privacy. The impact of the incentivisation models allows the creation of token-based incentive models, encouraging participation and collaboration within a network. The transforming capability fosters community engagement and loyalty, driving user involvement in decentralised ecosystems. The other distributed applications have an observable impact on the development of decentralised applications that operate on a peer-to-peer network. The transforming capabilities here are the development of decentralised applications that operate on a peer-to-peer network. The transformative capabilities of blockchain extend across industries, revolutionising traditional business models, increasing efficiency, and fostering a new era of trust and collaboration. The impact of blockchain continues to evolve as the technology matures and finds wider adoption.

The positive impacts that are mentioned above make blockchain technology a green technology. This is particularly because the efficiency in time, resources, and cost for a firm can lead to generating a competitive advantage. The competitive advantage of a firm is to ensure the customisation of the customer requirements in alignment with the wider stakeholder – the environment makes it eligible for being a green technology. The popularity of blockchain technology in the energy sector can emphasise that it is one of the most suitable technologies to be adopted to deliver net zero. The blockchain being implemented in the fast fashion industry is again another example of being the most suitable technology to achieve net zero in the long run.

The benefit of this study is targeted to the startups and entrepreneurs, environmentalists and sustainability advocates, non-profit organisations, legal professionals, builders and developers, technology enthusiasts and IT departments, educational institutions, healthcare providers, supply chain participants, government regulatory bodies, consumers, and most of the business corporations. The common benefits for a business corporation can be observed the streamline business processes, reduce costs, and improve efficiency by providing transparent and secure methods for record-keeping, supply chain management, and transactions. Consumers benefit from increased transparency, traceability, and authenticity of products. Blockchain can also empower consumers with more control over their personal data. Consumers benefit from increased transparency, traceability, and authenticity of products. Blockchain can also empower consumers with more control over their personal data. The participants in supply chains, including manufacturers, suppliers, and retailers, benefit from increased visibility, traceability, and efficiency. The healthcare sector can also benefit by implementing the blockchain by improving the security and interoperability of healthcare data, reducing fraud, and streamlining processes related to patient records and drug traceability. Though it is an element in the limitation section yet if the benefits are considered then the healthcare sector will benefit immensely from it. The new models and opportunities that blockchain opens for innovation, including the development of decentralised applications (DApps) and the creation of new business models have a very positive influence on the economy. Legal organisations will also reap fathomless benefits as blockchain technology can simplify and automate compliance processes, reduce the risk of fraud, and provide transparent records for legal purposes. In summary, blockchain technology has broad-reaching implications across diverse stakeholders, transforming the way business processes are conducted, data is managed, and trust is established in various sectors. The impact of blockchain continues to evolve as its adoption grows and new use cases emerge.

Some of the limitations of this study have been highlighted in this paragraph. The study here has a small scope including many other sectors like the disaster management sector, healthcare sector, and the banking sector. Disaster management has three parts, which are the pre-disaster or the planning phase, the response phase, and the recovery phase. These three are very critical aspects that can use blockchain technology to create a more lean and efficient humanitarian supply chain process. Due to lack of literature here, this aspect could not be studied. Again, similarly, the banking sector could not be studied here as the literature was not very relevant but the future research pathways can study how the implementation of the blockchain as a green technology can make the humanitarian supply chain, the banking sector, and the health sector more efficient for the wider stakeholders.

10.5 CONCLUSION

The structured literature review here has studied relevant literature that emphasises on the implementation of blockchain technology with the evolved features to understand the research question. The research question for this study is to investigate the implementation of blockchain technology as a green technology that can create sustainable sectors has been discussed with respect to its theoretical and practical implications. An answer to the research question is that the technology is capable of being green technology and can create sustainable sectors because of the knowledge that is shared with respect to its evolved features. Knowledge management becomes an important factor in understanding the opportunity to exploit the implementation at the right time. The blockchain technology has immense potential and can benefit a vast number of sectors. This transforming capability features based on knowledge sharing create sustainability.

REFERENCES

Aitzhan, N. Z., & Svetinovic, D. (2016). Security and privacy in decentralized energy trading through multi-signatures, blockchain and anonymous messaging streams. *IEEE Transactions on Dependable and Secure Computing, 15*(5), 840–852.

Aune, R. T., Krellenstein, A., O'Hara, M., & Slama, O. (2018). Footprints on a blockchain: Trading and information leakage in distributed ledgers. *The Journal of Trading (Retired), 13*(4), 49–57.

Barthélemy, M. (2011). Spatial networks. *Physics Reports, 499*(1–3), 1–101.

da Cruz, A. M. R., Santos, F., Mendes, P., & Cruz, E. F. (2020, May). Blockchain-based traceability of carbon footprint: A solidity smart contract for ethereum. In *ICEIS (2)*, Prague, Czechia. (pp. 258–268).

Dodge, E. T. (2015). A new model for carbon pricing using blockchain technology. https://www.edwardtdodge.com/2015/09/22/a-new-model-for-carbon-pricing-using-blockchain-technology/(Dodge, 2015)

du Can, S. D. I. R., & Price, L. (2008). Sectoral trends in global energy use and greenhouse gas emissions. *Energy Policy, 36*(4), 1386–1403.

EIA, U., & Kahan, A., (2019). EIA projects nearly 50% increase in world energy usage by 2050, led by growth in Asia. *Today in Energy*.

Energy Blockchain Labs (February, 2018). Available online: https://www.climate-kic.org/wp-content/uploads/2018/11/DLT-for-Climate-Action-Assessment-Nov-2018.pdf (accessed on 11 February 2018).

Fu, B., Shu, Z., & Liu, X. (2018). Blockchain enhanced emission trading framework in fashion apparel manufacturing industry. *Sustainability*, *10*(4), 1105.

Golding, O., Yu, G., Lu, Q., & Xu, X. (2022, May). Carboncoin: Blockchain tokenization of carbon emissions with ESG-based reputation. In *2022 IEEE International Conference on Blockchain and Cryptocurrency (ICBC)*, (pp. 1–5). Shanghai, China: IEEE.

Halu, A., Scala, A., Khiyami, A., & González, M. C. (2016). Data-driven modeling of solar-powered urban microgrids. *Science Advances*, *2*(1), e1500700.

Hong, T., Yan, D., D'Oca, S., & Chen, C. F. (2017). Ten questions concerning occupant behavior in buildings: The big picture. *Building and Environment*, *114*, 518–530.

Hwang, J., Choi, M. I., Lee, T., Jeon, S., Kim, S., Park, S., & Park, S. (2017). Energy prosumer business model using blockchain system to ensure transparency and safety. *Energy Procedia*, *141*, 194–198.

Jakobsson, M., & Juels, A. (1999, September). Proofs of work and bread pudding protocols. In *Secure Information Networks: Communications and Multimedia Security IFIP TC6/TC11 Joint Working Conference on Communications and Multimedia Security (CMS'99)*, September 20–21, 1999, Leuven, Belgium (pp. 258–272). Boston, MA: Springer US.

Khalilpour, R., & Vassallo, A. (2015). Leaving the grid: An ambition or a real choice? *Energy Policy*, *82*, 207–221.

Khaqqi, K. N., Sikorski, J. J., Hadinoto, K., & Kraft, M. (2018). Incorporating seller/buyer reputation-based system in blockchain-enabled emission trading application. *Applied Energy*, *209*, 8–19.

Lin, I. C., & Liao, T. C. (2017). A survey of blockchain security issues and challenges. *International Journal of Network Security*, *19*(5), 653–659.

Lin, Y. P., Petway, J. R., Anthony, J., Mukhtar, H., Liao, S. W., Chou, C. F., & Ho, Y. F. (2017). Blockchain: The evolutionary next step for ICT e-agriculture. *Environments*, *4*(3), 50.

Lu, J., Wu, S., Cheng, H., Song, B., & Xiang, Z. (2021). Smart contract for electricity transactions and charge settlements using blockchain. *Applied Stochastic Models in Business and Industry*, *37*(3), 442–453.

Ma, N., Waegel, A., Hakkarainen, M., Braham, W. W., Glass, L., & Aviv, D. (2023). Blockchain+ IoT sensor network to measure, evaluate and incentivize personal environmental accounting and efficient energy use in indoor spaces. *Applied Energy*, *332*, 120443.

Moazami, A., Nik, V. M., Carlucci, S., & Geving, S. (2019). Impacts of future weather data typology on building energy performance–Investigating long-term patterns of climate change and extreme weather conditions. *Applied Energy*, *238*, 696–720.

Mylrea, M., & Gourisetti, S. N. G. (2017, September). Blockchain for smart grid resilience: Exchanging distributed energy at speed, scale and security. In *2017 Resilience Week (RWS)* (pp. 18–23). Wilmington, DE, USA: IEEE.

Moher, D., Liberati, A., Tetzlaff, J., Altman, D.G. and PRISMA Group*, T., 2009. Preferred reporting items for systematic reviews and meta-analyses: the PRISMA statement. *Annals of Internal Medicine*, *151*(4), 264–269.

Pop, C., Cioara, T., Antal, M., Anghel, I., Salomie, I., & Bertoncini, M. (2018). Blockchain based decentralized management of demand response programs in smart energy grids. *Sensors*, *18*(1), 162.

Shah, J., Wollenberg, B. F., & Mohan, N. (2011, July). Decentralized power flow control for a smart micro-grid. In *2011 IEEE Power and Energy Society General Meeting* (pp. 1–6). Detroit, MI, USA: IEEE.

Sikorski, J. J., Haughton, J., & Kraft, M. (2017). Blockchain technology in the chemical industry: Machine-to-machine electricity market. *Applied Energy, 195*, 234–246.

Swan, M. (2015). *Blockchain: Blueprint for a New Economy*. O'Reilly Media, Inc.

Wood, G. (2014). Ethereum: A secure decentralised generalised transaction ledger. *Ethereum Project Yellow Paper, 151*(2014), 1–32.

Wörner, A., Meeuw, A., Ableitner, L., Wortmann, F., Schopfer, S., & Tiefenbeck, V. (2019). Trading solar energy within the neighbourhood: field implementation of a blockchain-based electricity market. *Energy Informatics, 2*, 1–12.

Part 5

Blockchain technology and practices in business management

Chapter 11

Blockchain applications in healthcare sector

Opportunities and challenges

Biswaksen Mishra and Rajesh Kumar Singh

11.1 INTRODUCTION

It is estimated by United Nations population division that world population is all set to reach 9.7 billion in 2050 and 11.2 billion by the year 2100. Global life expectancy at birth as of 2019 stands at 73.3 years up from 66.8 years in 2000. But the increase in life expectancy comes at the price of health deficiency. With ever degrading environment as well as the lifestyle choices, there is hardly any family which doesn't report any medical condition at any point in time (Science Daily, 2015). According to World Health Organization (2023), worldwide heart attack, stroke, lung cancer and chronic respiratory diseases are the fall-out of adverse climatic emissions and global warming. Infectious medical ailments such as HIV, tuberculosis, hepatitis, malaria and sexually transmitted diseases continue to afflict a wide population base.

So, essentially the demand for healthcare services is bound to see a rising curve over the decade. When viewed from the perspective of healthcare providers, there lies significant opportunity for tapping the potential market for healthcare sector. In the USA, healthcare is even considered a business domain where maximum financial benefits are exploited by the medical service providers from the interaction between the actors, i.e., doctors and patients (Sawyer, 2018). According to Gulf News report (2022), global home healthcare services are expected to grow at a compounded annual growth rate of 7.93% from 2022 to 2030. As per estimates of United Nations, with a geriatric population of 1.8 billion by 2060, accounting for 17.8% of the global population, demand for home healthcare service is believed to be tremendous (Gulf News report, 2022). However, with growing demand there are growing challenges as well. Globally, healthcare sector is haunted by a gamut of inefficient and unfair management practices. One of the common challenges in healthcare is the dispersed and decentralized medical health record of patients (Attaran, 2020; Dhagarra et al., 2020). In the absence of a consolidated and centralized medical history of a patient, proper or effective diagnosis and treatment of patients are difficult to achieve. Further, even if the medical records are made accessible in digital form, there is always a challenge to privacy and security of the records (Keshta & Odeh, 2021). Another challenge is inequity in healthcare services influenced by the factors of geographical access (Deloitte, 2023), insurance coverage (Barbosa & Cookson, 2019) and income (Barbosa & Cookson, 2019; Ilinca et al., 2019), among others. High cost of medicine and medical services is also a matter of serious concern for the consumers of healthcare sector (Perehudoff et al., 2016; Hammel & Michel, 2019). Healthcare sector is also afflicted with the issues of organ trafficking (Srivastava et al., 2021), sale of spurious or expired drugs (Clauson et al., 2018), doctor-pharma nexus and so on. On top of the above, there is severe shortage of human resources in the healthcare system (Marć et al., 2019), at a time of exploding population growth and increasing need for healthcare facilities.

However, many of the above issues can be addressed through digital technologies of Industry 4.0 (Cavallone & Palumbo, 2020). Blockchain technology is one such revolutionary technology of Industry 4.0 which can act to address the concerns of global medical healthcare system. Blockchain is a technology based on 'blocks' which contain records for every transaction or every new entry in digitized form and linked with each other through what is known as 'chains' through a process called mining (Attaran, 2020). Each block contains a cryptographic hash, a unique sequence of letters and numbers, linked to the data in the previous 'block' thereby establishing the 'chain' or network (Attaran, 2020; Bazel et al., 2021; Srivastava et al., 2021; Yaqoob et al., 2021; Zheng et al., 2017). Any change or updation or entry of new record gets created under a new block connected to its previous block through the cryptographic hash. So, the entire network or chain of blocks is an immutable one, resistant to tampering, alteration and so on (Attaran, 2020; Srivastava et al., 2021; Yaqoob et al., 2021). Blockchain, despite allowing decentralization of data, thereby sharing the data across all the blockchain partners like a democratic and resilient set-up (Anderson, 2019), is immutable to tampering and alteration thereby making it a secure and safe system of data management (Attaran, 2020; Agbo et al., 2019). Equipped with built-in features of decentralized storage, transparency, verifiability, immutability, ease of data access and others, blockchain technology finds immense usage in healthcare domains such as health record management, medical supply chain, medical insurance, disease diagnosis, medical supplies, medical treatment, drug and clinical trial and so on (Hasselgren et al., 2019; Xie et al., 2019; Srivastava et al., 2021; Ismail et al., 2021). The technology has the potential to solve the issue of data interoperability amongst the medical fraternity (Pirtle & Ehrenfeld, 2018). Srivastava et al. (2021) expect the blockchain technology in healthcare to grow at a CAGR of 8.7% from 2021–2026.

With unique characteristics of blockchain technology, making it easily accessible to the network partners of choice by the data owners under restricted viewing or usage while simultaneously offering unparalleled security and tampering owing to its technology features, application of blockchain in healthcare is the innovation of the upcoming decade which the academia and the medical community are looking up to.

Therefore, given the demand of healthcare expected to swell in the coming decade and the growth in blockchain applications predicted over the same period, it would be very interesting to study in further detail as to what are the challenges afflicting the healthcare industry and how is blockchain equipped to overcome those barriers and make the healthcare system efficient and effective for not just ordinary patients but also for the medical fraternity as a whole. To this effect, the current study intends to capture the following research questions:

RQ 1. What is the present trend of blockchain application research in healthcare?
RQ 2. What are the dominant challenges encountered in the healthcare system?
RQ 3. What opportunities do blockchain technology present to overcome the challenges of healthcare system?

The remaining part of the chapter is organized as follows: Section 11.2 discusses the methodology of the study. Section 11.3 presents the data analysis. Section 11.4 studies the challenges of healthcare system and the opportunities of using blockchain technology to address those challenges. Section 11.5 presents the conclusion of the study along with discussing some of the limitations and future scope of study.

11.2 METHODOLOGY

In order to identify the depth of discourse in the application of blockchain in healthcare and to address the challenges therein, a bibliometric analysis is conducted to explore emerging trends on the area of interest captured through intellectual discourse in academic journals (Donthu et al. 2021 Verma & Gustafsson, 2020). Bibliometric analysis is useful in uncovering meaningful cumulative scientific knowledge and evolutionary nuances from a collection of unstructured data in rigorous ways (Donthu et al., 2021). Thereafter, the important themes, as emerging from the literature, are discussed towards identification of the prominent challenges of the healthcare system and the ways blockchain addresses the same.

In order to identify the bibliographic collection for conducting bibliometric analysis, the relevant academic articles were chosen from SCOPUS database as it is considered to be a wide repository of academic literature and among the top two scientific databases considered for bibliometrics by the academic community presently (Donthu et al., 2021).

The articles were searched under SCOPUS database within Titles, Abstracts and Keywords using the search keywords ("blockchain" AND ("healthcare" OR "medical" OR "hospital") AND ("challenge" OR "opportunity")) restricted to document type as 'Articles' and subject area of "COMP" OR "ENGI" OR "MEDI" OR "DECI" OR "SOCI" OR "BUSI" OR "BIOC". The search results elicited 592 documents. Since the idea was to identify the potential applications and opportunities that blockchain provided for a healthcare supply chain, only those articles were included in the study which spoke only on the opportunities of blockchain technology for a healthcare context. Accordingly based upon scrutiny of the title and the abstract of the 592 documents resulting from the search outcome, 366 documents were identified for further study and rest were excluded. It included manuscripts accepted for publication in 2024 and the same were retained for further analysis, in order to obtain insights into the latest development in the topic of interest, given that the documents were equally reliable having been accepted after peer scrutiny.

In order to explore the discourse and the prominent themes of blockchain in healthcare, the bibliometric analysis was carried out over the said 366 documents using the software platform 'R'. The meta data of the aforesaid 366 documents were obtained from SCOPUS and the same were imported into the 'Biblioshiny' application of software 'R'. Running the 'Biblioshiny' application resulted in generating various synthesized information in the form of plots or graphs or tables. Some of the said information, found relevant for the study, was selected and is discussed in the next section.

11.3 DATA ANALYSIS

Results as obtained from the bibliometric analysis of the aforesaid documents using Biblioshiny application under the software platform 'R' are as described below.

11.3.1 Descriptive analysis

From the analysis of the bibliographic collection speaking on blockchain and healthcare (Figure 11.1), it is noticed that the relevant linkage has been identified from April 2017 onwards with annual rate of publication of literature on extant domains being 5.96% between April 2017 and August 2024. The entire such collection of documents is published under 193 academic journals. Aggregately, 1373 authors have contributed to the bibliographic set on the topic under study with only 11 authors having published articles under solo-authorship, while rest of the authors have published articles under co-authorship.

Blockchain applications in healthcare sector 221

Figure 11.1 Literature overview. (Bibliometrix R package.)

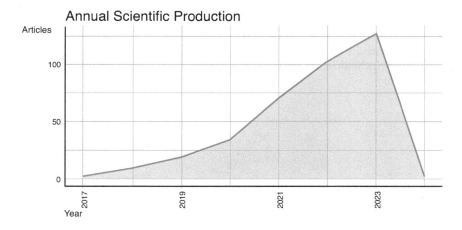

Figure 11.2 Annual scientific production of literature on blockchain and healthcare. (Bibliometrix R package.)

11.3.2 Academic penetration

Bibliometric analysis shows the significant growth of academic reports focusing on opportunities of blockchain in healthcare domain from 2017 to 2024, with the publications growing exponentially from 2020 to 2021 and continuing the growth in 2023 as well to touch a historical peak with more than 120 articles (127) on the topic published in the year 2023 till October, as can be seen from Figure 11.2 (Annual scientific production). This indicates the growing interest and importance of blockchain in healthcare within the academia so much so that two studies are already lined up for publication in 2024. So, it is very much implicative of the potential of blockchain application in healthcare sector and demand for research along this topic of relevance.

11.3.3 Co-occurrence network

Clear integration is visible in the academic discourse on blockchain application in healthcare sector as is evident from the co-occurrence network generated out of the bibliographic pool and depicted in Figure 11.3. From the figure, it is revealed that blockchain and healthcare clearly fall within a single cluster with close proximity between the nodes of blockchain and healthcare-related concepts such as healthcare or healthcare industry or healthcare system or medical services or hospitals, digital storage or hospital data processing or access control, electronic health record or medical record or health records or health data and more. Therefore, the above observation indicates that blockchain is seen as the potential technology to achieve better services and efficiency in healthcare domain. However, the plot also indicates that thus far majority of the studies see blockchain primarily as an efficient record maintenance tool for the healthcare sector, with some studies also discussing transaction related benefits due to the technology in healthcare sector. The second cluster appears to be focusing

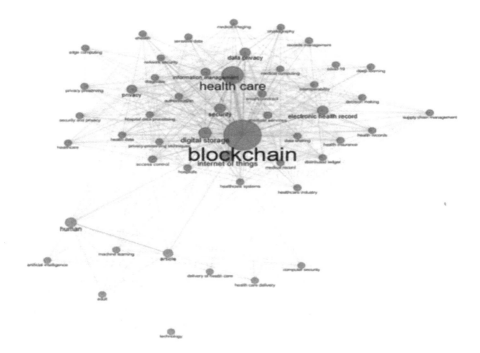

Figure 11.3 Co-occurrence network. (Bibliometrix R package.)

on the healthcare delivery while also discussing various technological interventions including blockchain, artificial intelligence and machine learning. So, though integration of blockchain and healthcare system is visible significantly in academic studies, there are few healthcare delivery specific studies as well where blockchain is yet to be explored fully.

11.3.4 Country scientific production

The prominent countries where the instant study is highly prominent among the academia are shown in Figure 11.4 which presents the percentage of the country of first author of the articles through a pie chart. India clearly stands out as the country from where maximum proportion of articles on blockchain applications in healthcare has been discussed followed by China. The USA features as the third country to produce literature and discussion on integration of blockchain with healthcare. This may be construed as the degree of penetration of the blockchain innovation or blockchain discourse in the countries with frequent and most publications, as conventionally any author is more interested or cognizant on domain of own country concern. Therefore, it could be said that India, one of the largest consumers of healthcare services and plagued with significant health related issues, is at the forefront of the discussion or implementation of blockchain in medical system, followed by China and the USA. Similar inferences could be drawn for other countries shown in the pie chart. As can be seen, India together with China, the USA and Saudi Arabia comprise nearly 50% of the contributors to the discussion on the blockchain in healthcare sector.

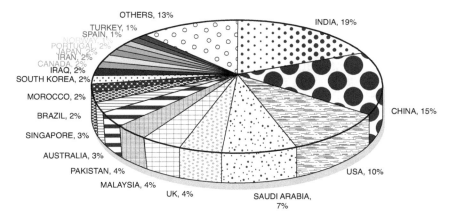

Figure 11.4 Percentage of articles produced by authors of respective countries (Generated by authors from SCOPUS results).

11.3.5 World cloud

The word cloud analysis of bibliographic collection, as obtained through Biblioshiny software, is plotted in Figure 11.5. This plot informs about the frequency of words featuring in the entire bibliographic pool, with greater the frequency of a word, the bigger its size and vice versa. Though the word 'blockchain' finds standout prominence in the literature pool, as is obvious given the selective choice of bibliographic material from the database followed by 'healthcare', what is important is the frequency of other words in the articles, suggestive of the importance and interconnection of the associated words featuring in the plot below. 'Internet of Things' is the next prominent word, which could be a reference to digital technologies including blockchain. 'Digital storage' features next reflecting the feasibility of digitization of health records courtesy of blockchain applications. Perhaps for the same reason, 'electronic health record', 'health record', 'medical record', 'hospital data processing', 'information management', 'record management' and similar words find a visible space in the word cloud, quite indicative of the potential attributed to blockchain application in maintaining electronic health records. 'Human' or 'Humans' is also seen in the word cloud with significant visibility since human beings are the sole subject in the healthcare sector being discussed. Among the other healthcare related terminologies, 'medical services', 'medical record', 'diagnosis', 'hospital data processing', 'healthcare delivery' or 'delivery of healthcare' also feature in the word cloud, though not in very significant proportions. Emergence of the said concepts reflects the association of blockchain as

Figure 11.5 Word cloud. (Bibliometrix R package.)

primarily an efficient records management technology. Other opportunities of blockchain within the healthcare system are yet to feature significantly in the academic research. This inference further corroborates the observation made under 'Co-occurrence network' according to which full-grown implementation or discussion of blockchain in healthcare is yet to come, though the process has already commenced.

11.4 CHALLENGES IN HEALTHCARE SYSTEM AND OPPORTUNITIES IN BLOCKCHAIN TECHNOLOGY

In order to further delve into some deeper nuances of the study on interconnection between healthcare and blockchain, it is important to discuss in detail the challenges grappling the healthcare system and understand why and how the application of blockchain in healthcare carries the potential to overcome those challenges. Accordingly, content analysis of forty articles out of the aforementioned bibliographic collection was undertaken, with broad consensual observations emerging out of the same, and have shaped our description as presented below.

11.4.1 Challenges in healthcare system

Despite the worldwide penetration of organized healthcare consumption, it is not bereft of challenges. Some of the prominent challenges in the healthcare system in the present decade are concerned with patient medical records or medical history, late diagnosis of vital life-threatening diseases and corruption among others. Each of the said challenges is described briefly as below.

11.4.1.1 Patient historical medical records

The most important challenge among all is related to patient medical records/history. There has hardly been any person in the universe who has been affected with just a one time or no medical condition throughout the lifetime. Almost all human beings suffer from multiple ailments as long as they live. And so, it is obvious that they create a medical history of their own and a series of medical reports associated with the ailments. But, it is quite seldom that the patient visits or treated by a single professional doctor who is well aware of the medical history of the patient. Rather the patient undergoes treatment across multiple hospitals under different doctors. So, in absence of any integrated medical data storage mechanism, the onus of apprising the medical history of patients to the attending doctor lies with the patient and his family and their ability to safe recordkeeping. Even then, the medical test reports have to be duplicated when moving from one hospital to another. Still, some hospitals even ask for fresh medical examination in disregard to the existing reports. This incurs cost and time to patients, apart from causing side effects due to exposure to harmful tests (Ismail et al., 2020; Bazel et al., 2021). Apart from the doctors, even the patients or their family members themselves and other stakeholders such as insurers, third parties, hospitals remain captivated to the fragmented medical data registered and stored across varied medical service providers (Dhagarra et al., 2020; Attaran, 2020). Even though this issue is being addressed with maintenance of electronic health records (EHR), the measure is far from giving it an integrated structure. Most of the hospitals maintain the medical history of patients under their treatment in electronic platform. Yet, such mechanism continues to remain in silo infrastructure. There is lack of integration or interoperability of the EHR amongst the medical system (Attaran, 2020). Data retrieval and sharing across other service providers in a smooth and secured manner remains a concern (Shahnaz et al., 2019; Dubovitskaya et al., 2020). Moreover, even with centralized cloud based system of medical data storage, data theft, hacking, privacy issues, interoperability of patient electronic health records across diverse medical service providers remain a stark reality (Ismail & Materwala, 2020; Yaqoob et al., 2021). According to Zarour et al. (2021), hacking of healthcare records and data breaching are on the rise. So, there is a dire need for an infrastructure providing secured yet integrated system of maintaining patient health records. Healthcare organization requires infrastructure with updated functionalities to maintain ECRs (Ismail et al., 2021).

11.4.1.2 Ineffective diagnosis of diseases

Absence of integrated medical records and inability of the patients and their family to present a comprehensive medical history before the professionals, that too in a timely manner, many a time leads to delayed or incorrect diagnosis of medical condition. Critical indicators of any vital or life-threatening disease such as cancer might stay hidden until the same is disclosed at an advanced stage (Lyratzopoulus et al., 2015). Late diagnosis of ill-health, especially life-threatening ones including COVID-19, not only risks medical casualty or permanent disability but is also more expensive as compared to early detection of symptoms (Fleishman et al. 2010; Varma et al., 2011; Oglesby et al., 2014; Baz et al.,2022; Salama & Eassa, 2022) where in most of the cases, barring a lucky few, it costs either life or money or any permanent disability to the patients and its inevitable side effects to their family members.

11.4.1.3 Corruption in healthcare

Healthcare system, like many other public welfare activities is not devoid of corruption. Major sphere of corruption observed in healthcare system is in organ trading and trafficking and circulation of counterfeit drugs among others. With advancement in technology, organ transplant has become a commonplace operation with doctors and medical system widely boasting of expertise in it. And so, with ever increasing diseases and ailments, there has been very high demand for organ transplants these days. Srivastava et al. (2021) states that there is a demand for organ transplant every 10 minutes, which makes it a very lucrative option for making money by means of organ trading and unfortunately for organ trafficking. Similarly, rising circulation of counterfeit drugs in the market is another area of concern (Sim et al., 2022). According to Kamel Boulus et al. (2018), authentic and quality drugs and medical supplies are one of the biggest challenges prevailing in the healthcare sector. As per Clauson et al. (2018), the size of counterfeit and spurious drugs globally valued $200 billion in the year 2016. The challenge is profound in developing or low-income countries (Srivastava et al., 2021). According to WHO, around 10 percent of medications sold around the world are believed to be fraudulent and the scope for entry of such spurious drugs into the system is through various stages of medical supply chain (Pathak et al., 2023). Another area which calls for some introspection in the health sector is insurance related frauds (Kapadiya et al., 2022). Studies have found a lot of insurance related frauds in health sector in the pre-blockchain era. Ismail and Zeadally (2021) discuss 12 types of health insurance frauds typically occurring at the level of doctors, hospitals, pharma companies, insurance providers and in certain scenario even involving the patients. Biggest concern is related to the manual documentation process for claim settlement which makes the detection of such frauds difficult to identify. Such frauds inflict severe loss of revenue to the exchequer of the concerned economies (Kapadiya et al., 2022).

11.4.2 Opportunities embedded in blockchain technology

All the above challenges can be tackled with the introduction of the revolutionary blockchain technology. Blockchain technology functions on primarily three permission management systems, namely, public, private and consortium (Ali et al., 2018; Bazel et al., 2021). In public blockchains, any individual connected to the internet can participate in the consensus procedure. Public blockchains integrate incentives and encrypted digit verification using proof-of-work or proof-of-stake mechanisms. The entire public blockchain system is transparent, wherein the identity of each participating individual remains pseudo-anonymous. In a private blockchain, only a single organization has control over the network. Therefore, such a type of blockchain requires a trustworthy agent to reach the consensus. The consortium blockchain combines the advantages of both public and private blockchain networks. It is only suitable for certain organizations that aim to streamline communication amongst one another (Yaqoob et al., 2021). Given the structure of blockchain technology and its features, adopting the technology in healthcare sector offers multiple opportunities, prominent of which are described below.

11.4.2.1 Electronic healthcare record

Blockchain technology enables data provenance related to medical health record of patients with full rights to the patients to control the accessibility of their data to other stakeholders and third parties as per consensus agreement built into the system (Srivastava et al., 2021; Yaqoob et al., 2021). All the information related to patients beginning from their initial medical test reports, prescriptions, treatment undertaken, medical facilities availed, details about healthcare providers, medical payment or billing history, patient's DNA data, biodata and others, collected from divergent sources such as laboratories, medical service providers, patient's data available from the service providers, smart devices, insurance providers and others, are stored on a single blockchain which gives a complete and comprehensive medical history and associated information at one glance (Attaran, 2020; Bazel et al., 2021; Srivastava et al., 2021; Yaqoob et al., 2021). Even the image data or sensor data are uploaded on the network seamlessly (Ali et al., 2023). Data of every patient is maintained uniquely through allocation of unique number based upon the patient's personal details, social security number, or financial details. As mentioned, the patient or any other network participants reserve the right to update, add any new data, and allow access to other interested parties, as the blockchain architecture decides for verification and usage of the visible data (Younis et al. 2021). The concept of cryptographic hash of blockchain ensures immutability, ensures integrity and safety of the records, prevents unauthorised access to the data, ease of data access and more (Wang et al., 2018; Ismail et al., 2021). Lifecode blockchain is a real example of permissioned blockchain adopted in the Chinese healthcare sector which allows restricted patient data sharing mechanism, with patients reserving the right to decide whom to share their personal data with (Jin et al., 2019). Patient data management system by Taipei Medical University and blockchain partner Digital Treasury Corporation also provide decentralised storage of data, with sharing control available with the patients, which not only ensures the security concerns of the data holders but also brings about transparency and efficiency in data usage (Jafri & Singh, 2022). Moreover, the infrastructure being built on a distributed ledger concept is stored and available across multiple stakeholders, which act as a backup and so is immune to any data loss due to any breakdown or disaster at any one of the nodes (Bazel et al., 2021; Yaqoob et al., 2021). Blockchain technology eliminates the need for any third party for data processing and maintenance, thereby reducing the associated administrative cost apart from ensuring non-disclosure of the personal data to third parties (Wang et al., 2018; Bazel et al., 2021; Ismail et al., 2021). By 2025, blockchain adoption is estimated to curb data breach related costs up to $100–150 bn per year (Yaqoob et al., 2021).

11.4.2.2 Early diagnosis and treatment of life-threatening diseases

Access to complete medical history at one glance courtesy blockchain technology enables better prognosis/diagnosis (Ismail et al., 2021; Bazel et al., 2021; Yaqoob et al., 2021). It aids the attending medical professionals to formulate or prescribe more accurate treatment plans or process (Srivastava et al., 2021). The data on patient medical history, treatment history, response trajectory, all available at a single platform from beginning to end also aids to address and optimize various clinical trial processes apart from ensuring clinical trial data authenticity and accuracy (Attaran, 2020; Yaqoob et al., 2021; Bazel et al., 2021). Tracing the effect of any drug on any patient as registered in the database of blockchain can not only disclose the allergic inclination of the patient but also help in accurate identification of the ailment. Moreover, it enables proactive treatment of genetic diseases and illnesses through genomic sequencing (Shivom, 2018). As per Yaqoob et al. (2021), 10% of chronic diseases are genetically inherited. Blockchain facilitates early detection of imminent diseases from primary symptoms with backtracked medical history and enables proactive treatment of root causes, thereby preventing disease from occurring in future (Attaran, 2020). This is especially helpful in contagious diseases such as COVID-19, wherein live locations of the incubators, real-time update of the health data would help in isolating the treatment and containing the spread of the disease (Baz et al., 2022). Similarly, the health data parameters extracted from many body wearable electronic devices introduced by technology companies and fitted in the blockchain data pertaining to any individual or human beings in general could be analysed for some insight into the lifestyle pattern and guiding on some preventive measures for better health and lifestyle (Attaran, 2020). Availability of comprehensive medical history online through blockchain also aids in effective teleconsultation without any data paucity issue.

11.4.2.3 Improved medical service offerings and customer empowerment

Another potential advantage of blockchain in healthcare is the scope of review of doctor's and hospital's performance (Pirtle & Ehrenfeld, 2018). With all relevant records pertaining to the academic and medical profile, professional performances of doctors available on the network, blockchain empowers patients to review and opt for the best medical service available amongst the given options based upon genuine data (Dewangan & Chandrakar, 2021). The same reviewing facility could be adopted for hospitals as well. This will not only save precious human lives from improper, apathetic and inefficient medical treatment but will also compel the medical fraternity to conduct their profession in a more efficient and sincere manner while serving as a deterrent for the inefficient, negligent and extortionist medical services. As an example, by tracing the effect of any drug on any patient as per the prescription of any doctor, it can be very well figured out at a later time whether the prescription offered was apposite or not. On the contrary, viewed from the service providers' perspective, blockchain offers the platform to the medical fraternity to showcase the best in the market medical services offered by them in terms of hospital facilities, clinics, doctors, laboratories and others to attract maximum patients. For patients, this information provides very crucial input in apt selection of the hospitals and medical facilities, not only in terms of the specialty of ailment but also in terms of spatial vicinity in emergency matters (Rejeb et al., 2019). Moreover, as discussed previously, blockchain helps in effective auditing of the healthcare system with availability of organized and objective data, which in turn ensures that the medical services provided by the hospitals and the attending medical professionals are efficient, effective and compliant with legal requirements (Bazel et al., 2021; Yaqoob et al., 2021). Because of the seamless data sharing option among the shareholders facilitated by blockchain, whether it is health record, image data or sensor data (Ali et al., 2023), it enables remote accessibility of all the shared data by any of the network participants or whosoever the data is shared with, from any part of the world (Chen, 2019) obviating the need for redundant multiple tests by any patient, thereby saving crucial time and money. Further, it facilitates patients or service providers to subscribe to remote and regular monitoring of health, especially for chronic diseases. (Azbeg et al., 2022).

11.4.2.4 Curbing corruption in healthcare

Blockchain technology is an extremely potential technology to arrest malafide activities in the medical system. Medical healthcare system is being increasingly plagued with issues of organ trafficking, circulation of fake drugs and alike. Blockchain technology helps in controlling organ trafficking (Hawashin et al., 2022), counterfeit drugs (Bazel et al., 2021, Attaran, 2020) and spurious medical devices (Gebreab et al., 2022). With features of data visibility across all the network partners and immutable data ledger, this technology leaves no scope for data tampering or hideous activities. Any malafide activity is discouraged as the source of such activity can be easily traced. Using blockchain, all the information related to availability of organs and its descriptives, donor data, recipient data, waiting list and organ health related information are stored in a time-stamped and cryptographically linked network visible to the donors, recipients and other concerned stakeholders (Hawashin et al., 2022). So, any activity to illegally trade in organs either by the hospital authorities, recipients, or donors, to the injustice of the affected party is curtailed from conception. Such a secure medical supply chain encourages acts of benevolent donations from healthy citizens as well as makes the system approachable and amiable for the needy potential recipients without any fear of exploitation. Such an improved system could also facilitate in establishing resource bank related to medical donations in organs, blood, tissues and so on. The technology is also immensely helpful in restricting drug trafficking or circulation of spurious drugs as the cryptographic chain captures the data across all stages of the drug supply chain initiating from the manufacturing of drugs till the final purchase by consumers (Kumar & Tripathi, 2019; Sahoo et al., 2020; Mazlan et al., 2020; Yazdinejad et al., 2023). This also empowers the consumers to authenticate the drugs at the time of their purchase (Haq & Muselemu, 2018). Even the customers can identify the true validity of the drugs, rather than falling for spurious drugs with identical make-over and/or re-designed fictitious validity period for consumption of the products (Srivastava et al., 2021). FarmaTrust, a blockchain based solution for pharmaceutical supply chain, claims to eliminate the issue of counterfeit drugs by offering end-to-end visibility and transparency of data across the pharmaceutical supply chain.[1] Another similar application is the eZTracker, blockchain solution by Zuellig Pharma, which provides end-to-end data connectivity and provenance for the pharma supply chain, thereby identifying more than 6700 potential counterfeits and illegal trade of medical products circulating in the market of Hong Kong and Thailand (Sim et al., 2022).

11.4.2.5 Improved efficiency in medical supply chain

Due to absence of an automated system to monitor, track and replenish medical supplies, many a time, there is demand–supply mismatch which disrupts the chain profoundly. Blockchain has the potential to facilitate easy monitoring and tracking of the movement of medical supplies along the supply chain (Abdallah & Nizamuddin, 2023) which can play most vital role in accessibility of medicine to the vulnerable section of the society at the time of most requirement. Most recent example in sight being the scenario during COVID-19 when there was significant supply shortfall in medicines, medical equipment and all other accessories globally. Such situation can be averted through a blockchain enabled medical supply chain by tracking the demand and supply chain (Baz et al., 2022). eZTracker is one such application which triggers auto-replenishment of medical supplies at clinic or hospital level based on the real-time inventory position (Laverick, 2022). Further, in a situation of severe demand in a pre-blockchain era, cases of black-marketing of spurious drugs, inferior quality of medical equipment and accessories are not unheard of. Using blockchain technology, such an eventuality could be mitigated to a lot extent in multiple ways (Liu et al., 2024). Firstly, by identification of the demand pockets and its requirement from the ledger data in the block, supplies could be channelized as per requirement. Secondly, due to the ease of instant data visibility of the demand indents in the blocks, it provides an outlook for demand not only at local level but also indicates the geographical demand clusters and hence enables the manufacturer to track its fast depleting stocks and to take early actions to meet the requirement. Thirdly, due to the visibility across the network partners, the customers may also track the movement of the drugs and remain more informed on the availability of the drug, thereby refraining from falling into the enticement of black-marketing of the products. Further, in the absence of any intermediary, accuracy of data, feasibility of real-time data tracking, blockchain is able to bring about efficiency in the pharma supply chain (Bandhu et al., 2023). Moreover, as has been mentioned previously, due to immutability of the data in the blocks, no artificial supply shortfall or demand shortfall can be manufactured by vested intermediaries hooked into the network. Talking about quality, the devices or the drugs require standardized environmental conditions during shipment. As such, by analysing the said parameters in the blocks, the quality consistency can be tracked from the start till the end (Srivastava et al., 2021).

11.4.2.6 Efficiency in medical insurance segment

Insurance domain of healthcare system is another big beneficiary of blockchain technology. Through blockchain technology, the tedious, time-consuming and inefficient process of sharing vital patient and health insurance related information across all the stakeholders is eliminated (Yaqoob et al., 2021; Velliangiri et al., 2023). As mentioned previously, manual insurance claim settlement process is a big contributor to insurance related frauds (Ismail & Zeadally, 2021; Kapadiya et al., 2022). It makes detection of fraud claims a tedious and difficult task. However, by making the data available on digital platform and that too with end-to-end availability of medical history, auditing the claim becomes very transparent, smooth and efficient. However, many channel partners such as hospitals or insurance providers and even customers are quite averse to sharing data online over privacy concerns (Velliangiri et al., 2023). By ensuring data privacy and security over the digital platform, yet making the data accessible to the concerned nodes in a restricted manner, blockchain facilitates efficient claim lodging and settlement process for the healthcare sector (Kapadiya et al. 2022; Sutanto et al., 2022). According to Giancaspro (2017), smart contracts are self-executing contracts with generic terms and conditions between the insurer and insurance provider penned in codes and shared with restricted access across the blockchain network partners. Smart contracts in a blockchain allow automatic collection of the relevant data and information. The policy renewals and claims settlement processes can also be very quick due to smart contracts (Ismail & Zeadally, 2021). Further, due to better and accurate registration of patient history, the insurance premium is decided more pragmatically and averts any future claim related disputes due to masked medical conditions as occur during traditional medical insurance cases. Moreover, smart contact is a very powerful tool to process insurance related payment and billing. Many a time, due to medical treatment over wide assortment of specialty hospitals according to medical condition or due to some other concern, the processing of insurance claims gets entangled in paperworks or other process issues (Ismail & Zeadally, 2021). Even the billing process due to blockchain is much improved, transparent and quick, resulting in quick settlement of claims (Bazel et al., 2021; Yaqoob et al., 2021).

11.5 CONCLUSION, LIMITATIONS AND FUTURE SCOPE

Healthcare industry is set to boom with population growth. With diverse toxic and polluting environmental factors having penetrated deep into the society with ever increasing pace, health of living beings has become easy target of contamination. New variants of diseases are diffusing into society, whether air-borne, water-borne or vector-borne. At such a crucial juncture, it is important to reflect on the challenges afflicting the healthcare industry and bring in technological intervention that is able to address the issues of healthcare sector head-on. The issues of electronic health record, untimely, delayed or erroneous diagnosis of ailment, corruption, medical supply chain inefficiencies have spread their tentacles in and out of the industry. Blockchain is one such technology of the future which can bring about process innovations to fix challenges of healthcare system apart from unsettling other parallel industrial structures. Blockchain is a multi-faceted tool equipped to tackle multiple processual vices. From the perspective of operations management, this revolutionary technology can be safely considered one of the most effective quality tools for bringing in process efficiency in healthcare system. The dominant strength of blockchain technology is its immutability to tampering, restricted accessibility, shared visibility through distributed ledger, data history tracking ability, easy new data registration, ease of searching and linkage of past data, remote accessibility and so on. On the above strengths, maintenance of electronic health record, early and effective diagnosis of diseases, metrics for performance of medical professionals, medical services and medical infrastructure, inhibiting circulation of counterfeit drugs, preventing trafficking in organs and drugs, effective medical supply chain performance and efficient medical insurance services can be attained in healthcare system in future. For transpiring the above process improvement, blockchain technology must find its way into the philosophy and strategy of global medical system.

Despite conducting the study in a fair and rigorous manner to discuss the challenges in healthcare sector and opportunities through application of blockchain, the study has some limitations which may provide further scope of study along this domain. The study has been conducted based on the academic literature available in renowned database. However, it is worth conducting the study by adopting inductive method by getting some qualitative interviews from the practitioners, in the process of or having already implemented blockchain in their service delivery, whether in healthcare sector or otherwise, to get some practical perspective on the innovation. Secondly, the study has been conducted as a global concern and the challenges and opportunities described are based upon the generalized ones. Future study could also focus on specific geography or socio-economic classification or institutional classification or any other category to introspect on specific challenges relevant to those sectors and discuss the application of blockchain to address them. Thirdly, this chapter has endeavoured to advocate the opportunities of blockchain innovation in healthcare. But the challenges which blockchain in health sector is likely to pose may be explored further. Given that digital penetration and digital literacy are yet to be universalized, especially at the lower strata of the society, there might be some challenges which the blockchain technology may pose as an interface between the consumer and the medical service provider. Further, apart from the direct stakeholders in the value chain, there might be a lot of indirect stakeholders in the chain as well, such as state institutions or some third-party financial intermediaries or others. What role do they play in seamless integration of blockchain may also be seen? Similarly, the role of foreign stakeholders in the value chain and how it is regulated under any state or national policies may also be explored in future.

Nevertheless, infusion of blockchain technology in healthcare appears imminent in the near future. As such, as a society we must be fully prepared to adapt to the new technology, if we wish to harness the opportunity the technology is capable of offering to make the global healthcare system much more efficient, patient-friendly, industry-friendly and devoid of existing challenges.

NOTE

1 https://www.farmatrust.com/copy-of-pharmaceutical-tracking-dat.

REFERENCES

Abdallah, S., & Nizamuddin, N. (2023). Blockchain-based solution for pharma supply chain industry. *Computers & Industrial Engineering*, 177, 108997.

Agbo, C., Mahmoud, Q., & Eklund, J. (2019). Blockchain technology in healthcare: A systematic review. *Healthcare*, 7, 56. https://doi.org/10.3390/healthcare7020056.

Ali, M. S., Vecchio, M., Pincheira Caro, M. R., Dolui, K., Antonelli, F., & Rehmani, M. H. (2018). Applications of blockchains in the internet of things: A comprehensive survey. *IEEE Communications Surveys & Tutorials*, 1–1. https://doi.org/10.1109/COMST.2018.2886932.

Ali, A., Ali, H., Saeed, A., Ahmed Khan, A., Tin, T. T., Assam, M., ... & Mohamed, H. G. (2023). Blockchain-powered healthcare systems: Enhancing scalability and security with hybrid deep learning. *Sensors*, 23(18), 7740.

Anderson, M. (2019). Exploring decentralization: Blockchain technology and complex coordination. *Journal of Design and Science*. https://jods.mitpress.mit.edu/pub/7vxemtm3

Attaran, M. (2020). Blockchain technology in healthcare: Challenges and opportunities. *International Journal of Healthcare Management*. 1–15. https://doi.org/10.1080/20479700.2020.1843887.

Azbeg, K., Ouchetto, O., & Andaloussi, S. J. (2022). Access Control and Privacy-Preserving Blockchain-Based system for diseases management. *IEEE Transactions on Computational Social Systems*, 10(4), 1515–1527.

Bandhu, K. C., Litoriya, R., Lowanshi, P., Jindal, M., Chouhan, L., & Jain, S. (2023). Making drug supply chain secure traceable and efficient: a Blockchain and smart contract based implementation. *Multimedia Tools and Applications*, 82(15), 23541–23568.

Barbosa, E. C., & Cookson, R. (2019). Multiple inequity in health care: An example from Brazil. *Social Science & Medicine*, 228, 1–8.

Baz, M., Khatri, S., Baz, A., Alhakami, H., Agrawal, A., & Khan, R. A. (2022). Blockchain and artificial intelligence applications to defeat COVID-19 pandemic. *Computer Systems Science & Engineering*, 40(2), 691–702.

Bazel, M., Mohammed, F., & Ahmad, M. (2021). Blockchain technology in healthcare big data management: Benefits, *Applications and Challenges*, 1–8. https://doi.org/10.1109/eSmarTA52612.2021.9515747.

Cavallone, M., & Palumbo, R. (2020). Debunking the myth of industry 4.0 in health care: insights from a systematic literature review. *The TQM Journal*, 32(4), 849–868.

Chen, H. S., Jarrell, J. T., Carpenter, K. A., Cohen, D. S., & Huang, X. (2019). Blockchain in healthcare: a patient-centered model. *Biomedical Journal of Scientific & Technical Research*, 20(3), 15017.

Clauson, K., Breeden, E., Davidson, C., & Mackey, T. (2018). Leveraging blockchain technology to enhance supply chain management in healthcare. *Blockchain in Healthcare Today*. https://doi.org/10.30953/bhty.v1.20.

Deloitte (2023). 2023 Global Health Care Outlook. https://www2.deloitte.com/cn/en/pages/life-sciences-and-healthcare/articles/2023-global-health-care-outlook.html.

Dewangan, N. K., & Chandrakar, P. (2021, December). Patient feedback based physician selection in blockchain healthcare using deep learning. In *International Conference on Advanced Network Technologies and Intelligent Computing* (pp. 215–228). Cham: Springer International Publishing.

Dhagarra, D., Goswami, M., & Kumar, G. (2020). Impact of trust and privacy concerns on technology acceptance in healthcare: An Indian perspective. *International Journal of Medical Informatics*, 141, 104164. https://doi.org/10.1016/j.ijmedinf.2020.104164.

Donthu, N., Kumar, S., Mukherjee, D., Pandey, N., & Lim, W. M. (2021). How to conduct a bibliometric analysis: An overview and guidelines. *Journal of Business Research*, 133, 285–296.

Dubovitskaya, A., Baig, F., Xu, Z., Shukla, R., Zambani, P. S., Swaminathan, A., ... & Wang, F. (2020). ACTION-EHR: Patient-centric blockchain-based electronic health record data management for cancer care. *Journal of medical Internet research*, 22(8), e13598.

Fleishman, J. A., Yehia, B. R., Moore, R. D., Gebo, K. A., & HIV Research Network. (2010). The economic burden of late entry into medical care for patients with HIV infection. *Medical Care*, 48(12), 1071.

Gebreab, S. A., Hasan, H. R., Salah, K., & Jayaraman, R. (2022). NFT-based traceability and ownership management of medical devices. *IEEE Access*, 10, 126394–126411.

Giancaspro, M. (2017). Is a 'smart contract' really a smart idea? Insights from a legal perspective. *Computer Law & Security Review*. 33. https://doi.org/10.1016/j.clsr.2017.05.007.

Gulf News Report. (2022). Home Healthcare Services sector is set for steady growth in the next decade. *Corporate-news – Gulf News*. Available at https://gulfnews.com/business/corporate-news/home-healthcare-services-sector-is-set-for-steady-growth-in-the-next-decade-1.1666794157812#:~:text=The%20global%20home%20healthcare%20services%20are%20expected%20to,diseases%20are%20factors%20expected%20to%20fuel%20industry%20growth.

Hammel, B., & Michel, M. C. (2019). Why are new drugs expensive and how can they stay affordable? *Concepts and Principles of Pharmacology: 100 Years of the Handbook of Experimental Pharmacology*, 453–466.

Haq, I. & Muselemu, O. (2018). Blockchain technology in pharmaceutical industry to prevent counterfeit drugs. *International Journal of Computer Applications*, 180, 8–12. https://doi.org/10.5120/ijca2018916579.

Hasselgren, A., Kralevska, K., Gligoroski, D., Pedersen, S., & Faxvaag, A. (2019). Blockchain in healthcare and health sciences-A scoping review. *International Journal of Medical Informatics*, 134, 104040. https://doi.org/10.1016/j.ijmedinf.2019.104040.

Hawashin, D., Jayaraman, R., Salah, K., Yaqoob, I., Simsekler, M. C. E., & Ellahham, S. (2022). Blockchain-based management for organ donation and transplantation. *IEEE Access*, 10, 59013–59025.

Ilinca, S., Di Giorgio, L., Salari, P., & Chuma, J. (2019). Socio-economic inequality and inequity in use of health care services in Kenya: evidence from the fourth Kenya household health expenditure and utilization survey. *International Journal for Equity in Health*, 18, 1–13.

Ismail, L., & Zeadally, S. (2021). Healthcare insurance frauds: Taxonomy and blockchain-based detection framework (Block-HI). *IT professional*, 23(4), 36–43.

Ismail, L. & Materwala, H. (2020). BlockHR: A blockchain-based framework for health records management, 164–168. https://doi.org/10.1145/3408066.3408106.

Ismail, L., Materwala, H., & Hennebelle, A. (2021). A scoping review of integrated blockchain-cloud (BcC) architecture for healthcare: Applications, challenges and solutions. *Sensors*, 21, 3753. https://doi.org/10.3390/s21113753.

Ismail, L., Materwala, H., & Khan, M. (2020). Performance evaluation of a patient-centric blockchain-based healthcare records management framework, 39–50. https://doi.org/10.1145/3409934.3409941.

Jafri, R., & Singh, S. (2022). Blockchain applications for the healthcare sector: Uses beyond Bitcoin. *Blockchain Applications for Healthcare Informatics*, 71–92.

Jin, X. L., Zhang, M., Zhou, Z., & Yu, X. (2019). Application of a blockchain platform to manage and secure personal genomic data: a case study of LifeCODE. AI in China. *Journal of medical Internet research*, 21(9), e13587.

Kamel Boulos, M., Wilson, J., & Clauson, K. (2018). Geospatial blockchain: Promises, challenges, and scenarios in health and healthcare. *International Journal of Health Geographics*, 17. https://doi.org/10.1186/s12942-018-0144-x.

Kapadiya, K., Patel, U., Gupta, R., Alshehri, M. D., Tanwar, S., Sharma, G., & Bokoro, P. N. (2022). Blockchain and AI-empowered healthcare insurance fraud detection: An analysis, architecture, and future prospects. *IEEE Access*, 10, 79606–79627.

Keshta, I., & Odeh, A. (2021). Security and privacy of electronic health records: Concerns and challenges. *Egyptian Informatics Journal*, 22(2), 177–183.

Kumar, R. & Tripathi, R. (2019). Traceability of counterfeit medicine supply chain through Blockchain, 568–570. https://doi.org/10.1109/COMSNETS.2019.8711418.

Laverick, D. (2022). Life science business value of blockchain case use blockchain real world deployment taking bad actors and counterfeit drugs off the market: Blockchain real world deployment taking bad actors and counterfeit drugs off the market. *Blockchain in Healthcare Today*, 5 doi: https://doi.org/10.30953/bhty.v5.203

Liu, X., Shah, R., Shandilya, A., Shah, M., & Pandya, A. (2024). A systematic study on integrating blockchain in healthcare for electronic health record management and tracking medical supplies. *Journal of Cleaner Production*, 447, 141371.

Lyratzopoulos, G., Vedsted, P., & Singh, H. (2015). Understanding missed opportunities for more timely diagnosis of cancer in symptomatic patients after presentation. *British Journal of Cancer*, 112(1), S84–S91.

Marć, M., Bartosiewicz, A., Burzyńska, J., Chmiel, Z., & Januszewicz, P. (2019). A nursing shortage–a prospect of global and local policies. *International Nursing Review*, 66(1), 9–16.

Mazlan, A., Mohd Daud, S., Mohd Sam, S., Abas, H., Rasid, S., & Yusof, F. (2020). Scalability challenges in healthcare blockchain system — A systematic review. *IEEE Access*, 8. 1–1. https://doi.org/10.1109/ACCESS.2020.2969230.

Oglesby, A., Korves, C., Laliberté, F., Dennis, G., Rao, S., Suthoff, E. D., ... & Duh, M. S. (2014). Impact of early versus late systemic lupus erythematosus diagnosis on clinical and economic outcomes. *Applied Health Economics and Health Policy*, 12, 179–190.

Pathak, R., Gaur, V., Sankrityayan, H., & Gogtay, J. (2023). Tackling counterfeit drugs: The challenges and possibilities. *Pharmaceutical Medicine*, 1–10.

Perehudoff, K., Toebes, B., & Hogerzeil, H. (2016). A human rights-based approach to the reimbursement of expensive medicines. *Bulletin of the World Health Organization*, 94(12), 935.

Pirtle, Claude & Ehrenfeld, Jesse. (2018). Blockchain for healthcare: The next generation of medical records? *Journal of Medical Systems*, 42, 172. https://doi.org/10.1007/s10916-018-1025-3.

Rejeb, A., Keogh, J. G., & Treiblmaier, H. (2019, December). The impact of blockchain on medical tourism. In *Workshop on E-Business* (pp. 29–40). Cham: Springer International Publishing.

Sahoo, M., Samanta Singhar, S., & Sahoo, S. (2020). A blockchain based model to eliminate drug counterfeiting. https://doi.org/10.1007/978-981-15-1884-3_20.

Salama, A. S., & Eassa, A. M. (2022). IOT and cloud based blockchain model for COVID-19 infection spread control. *Journal of Theoretical and Applied Information Technology*, 100(1), 113–126.

Sawyer, N. T. (2018). In the U.S. "Healthcare" is now strictly a business term. *Western Journal of Emergency*. 19(3), 494. doi.org: 10.5811/westjem.2018.1.37540.

Science Daily (2015). Over 95% of the world's population has health problems, with over a third having more than five ailments. Available at https://www.sciencedaily.com/releases/2015/06/150608081753.htm.

Shahnaz, A., Qamar, U., & Khalid, A. (2019). Using blockchain for electronic health records. *IEEE Access*, 1–1. https://doi.org/10.1109/ACCESS.2019.2946373.

Shivom. (2018). Blockchain can be the catalyst for a revolution in precision medicine. Medium. https://medium.com/projectshivom/blockchain-can-be-the-catalyst-for-a-revolution-in-precision-medicine-d55e1e810262

Sim, C., Zhang, H., & Chang, M. L. (2022). Improving end-to-end traceability and pharma supply chain resilience using blockchain. *Blockchain in Healthcare Today*, 5, doi: 10.30953/bhty.v5.231.

Srivastava, V., Mahara, T., & Yadav, P. (2021). An analysis of the ethical challenges of blockchain-enabled e-healthcare applications in 6G networks. *International Journal of Cognitive Computing in Engineering*, 2. https://doi.org/10.1016/j.ijcce.2021.10.002.

Sutanto, E., Mulyana, R., Arisgraha, F. C. S., & Escrivá-Escrivá, G. (2022). Integrating blockchain for health insurance in Indonesia with hash authentication. *Journal of Theoretical and Applied Electronic Commerce Research*, 17(4), 1602–1615.

Varma, R., Lee, P. P., Goldberg, I., & Kotak, S. (2011). An assessment of the health and economic burdens of glaucoma. *American Journal of Ophthalmology*, 152(4), 515–522.

Velliangiri, S., Karthikeyan, P., Ravi, V., Almeshari, M., & Alzamil, Y. (2023). Intelligence amplification-based smart health record chain for enterprise management system. *Information*, 14(5), 284.

Verma, S., & Gustafsson, A. (2020). Investigating the emerging COVID-19 research trends in the field of business and management: A bibliometric analysis approach. *Journal of Business Research*, 118, 253–261.

Wang, S., Wang, J., Wang, X., Qiu, T., Yuan, Y., Ouyang, L., & Guo, Y. (2018). Blockchain-powered parallel healthcare systems based on the ACP approach. *IEEE Transactions on Computational Social Systems*, 1–9. https://doi.org/10.1109/TCSS.2018.2865526.

World Health Organization (2023). Climate change and noncommunicable diseases: connections. Available at https://www.who.int/news/item/02-11-2023-climate-change-and-noncommunicable-diseases-connections.

Xie, J., Tang, H., Huang, T., Yu, F., Xie, R., Liu, J., & Liu, Y. (2019). A survey of blockchain technology applied to smart cities: Research issues and challenges. *IEEE Communications Surveys & Tutorials*, 1–1. https://doi.org/10.1109/COMST.2019.2899617.

Yaqoob, I., Salah, K., Jayaraman, R., & Al-Hammadi, Y. (2021). Blockchain for healthcare data management: Opportunities, challenges, and future recommendations. *Neural Computing and Applications*. 34. https://doi.org/10.1007/s00521-020-05519-w.

Yazdinejad, A., Rabieinejad, E., Hasani, T., & Srivastava, G. (2023). A BERT-based recommender system for secure blockchain-based cyber physical drug supply chain management. *Cluster Computing, 26,* 3389–3403.

Younis, M., Lalouani, W., Lasla, N., Emokpae, L., & Abdallah, M. (2021). Blockchain-enabled and data-driven smart healthcare solution for secure and privacy-preserving data access. *IEEE Systems Journal,* 16(3), 3746–3757.

Zarour, M., Alenezi, M., Ansari, M. T. J., Pandey, A. K., Ahmad, M., Agrawal, A., ... & Khan, R. A. (2021). Ensuring data integrity of healthcare information in the era of digital health. *Healthcare Technology Letters,* 8(3), 66–77.

Zheng, Z., Xie, S., Dai, H.-N., Chen, X., & Wang, H. (2017). An overview of blockchain technology: Architecture, consensus, and future trends. https://doi.org/10.1109/BigDataCongress.2017.85.

Chapter 12

Improving transparency and efficiency in international trade through blockchain technology

Witold Bahr, Lee Ee Yern, and Ian McEwan

12.1 INTRODUCTION

The inception of Bitcoin and the underlying blockchain technology has since established a system for digital transactions between two parties, predicated on cryptographic proof, thereby eliminating the need for an external financial institution (Nakamoto, 2008). As per the definition provided by the United States Sentencing Commission (USSC) (2018), a blockchain is a tamper-resistant distributed ledger, safeguarded by cryptography, that maintains a record of all transactions occurring on the network and ensures transparency for all participants. These attributes render blockchain technology apt for industries necessitating secure, expedient, and transparent transactions, such as banking and finance (Saheb & Mamaghani, 2021), supply chains (Saberi et al., 2019), and healthcare (Samsir & Jain, 2022), among others.

A compelling rationale for the adoption of blockchain technology in supply chains, especially in the context of international trade, is the potential to diminish or even eradicate paperwork, thereby mitigating administrative expenses, fraud, and human errors. This paperwork reduction extends to various aspects of the supply chain, such as trade financing and customs clearance, facilitated by the integration of smart contracts and the Internet of Things (IoT) (Ganne, 2018).

In this context, trade financing pertains to bank-mediated trade finance, where banks facilitate transactions between purchasers and sellers. Historically, international trade has relied on trade financing due to the inherent risks, which are greater than those in domestic trade. Consequently, these risks are assumed by banks rather than the parties directly involved in the trade (Serena & Vasishtha, 2015). A multiple-case study demonstrated that the application of consortium blockchains, coupled with smart contracts for trade financing, enhanced process automation, traceability, and reliability for the involved parties. Furthermore, it led to cost reductions and incentivised trade participants to maintain accountability, as all transactions were recorded on a shared ledger (Chang et al., 2019b).

The customs clearance procedure has been pinpointed as a substantial impediment in the supply chain, primarily due to its dependence on paper-based processes. This system, which heavily relies on physical documentation, has led to a disconnect between analogue and digital data, an unnecessary duplication of information flow, significant inaccuracies in registration data, and communication inefficiencies upon process completion (Boschian et al., 2010). Consequently, the incorporation of blockchain technology can augment the customs clearance procedure by enhancing data transparency and accessibility for all stakeholders, thereby ensuring better tax compliance and mitigating financial crimes (Okazaki, 2018).

Blockchain technology, being in its nascent phase, will require a substantial period for its assimilation, predominantly due to the absence of regulatory mechanisms. A significant proportion of policymakers are inadequately informed about blockchain, which poses a hurdle to the incorporation of blockchain-centric systems in the broader industry (Norberg, 2019). Additional obstacles, such as technical interoperability between blockchain systems and current digital systems, data standardisation, scalability, and apprehensions related to energy consumption, persist as areas for further scrutiny (Ganne, 2018).

The objectives of this chapter are manifold. Firstly, it aims to scrutinise the existing body of research to understand the application, advantages, and challenges of blockchain technology in international trade. In doing so, it addresses the research gap in the current literature, which, out of thousands of papers on blockchain, contains just over forty considering its implications for international trade, and summarises findings in a convenient framework (see Section 12.3). Secondly, the novelty of this chapter is exemplified through its practical application via a case study. This case study not only provides a real-world context but also contributes to the academic discourse by offering fresh insights into the implementation of blockchain technology. Thus, this chapter contributes to the field by bridging the research gap and enhancing the understanding of the potential benefits and challenges of integrating blockchain technology into international trade and broader supply chain operations. As such, this chapter addresses the following questions:

1. What are the applications of blockchain technology in the sphere of international trade?
2. What are the advantages offered by blockchain technology in international trade?
3. What are the hurdles encountered in the application of blockchain technology in international trade?

Section 12.2 provides an overview of key aspects in international trade, encompassing supply chains, trade finance, and customs clearance.

It explores the challenges associated with these elements and introduces blockchain technology, delving into both its advantages and the challenges linked to its integration into international trade practices. Moving forward, Section 12.3 engages in a comprehensive discussion regarding the applications, benefits, and challenges posed by blockchain in the realm of international trade. Section 12.4 presents a case study showcasing blockchain in international trade, demonstrating its practical use. Lastly, Section 12.5 concludes with a brief summary of the findings.

12.2 BACKGROUND TO INTERNATIONAL TRADE AND BLOCKCHAIN

International trade is characterised as an economic exchange between nations, typically involving consumer and capital goods (Robinson et al., 2020). The explosion of international trade was due to globalisation, which enabled the growth of exports from underdeveloped countries (Gereffi, 1999). As of 2020, international seaborne trade has expanded by a factor of 18.5 times since 1980 (Statista, 2020).

The evidence indicates a strong positive correlation between a country's level of trade and its GDP per capita, suggesting that trade can be a significant catalyst for economic growth and development (Irwin & Terviö, 2002). A crucial element that facilitated this was the rise of information and communication technology (ICT), which eliminated geographic boundaries between countries and individuals, enabling international trade through globalisation (Latif et al., 2018). It was observed that developing economies that embraced globalisation and engaged in international trade experienced an increase in their production output (Dodzin & Vamvakidis, 2004).

The international trade supply chain encompasses several key processes, including customs clearance, trade finance, logistics, and trading (Chang et al., 2019a).

12.2.1 Customs clearance

Customs clearance, which facilitates trade across countries, is tasked with ensuring the legitimacy, safety, and security of goods (Petersen et al., 2018). It has been demonstrated to have a statistically significant correlation with import and export volume, indicating that maintaining efficiency in this area is crucial for successful international trade (Gani, 2015). ICT has been a pivotal driver in enhancing the efficiency of customs clearance. It has enabled instantaneous data processing at each stage of the customs clearance process, reducing reliance on "paper shuffling" processes and decreasing communication time between relevant parties (Appeals & Struye de Swielande, 1998; Li and Li, 2019).

12.2.2 Trade finance

Trade finance is characterised as any banking product or service that aids international trade by facilitating risk management and providing working capital (Committee on the Global Financial System, 2014). It is also utilised by trading parties to distribute risk and protect against fraud (Kowalski et al., 2021). The four main types of trade finance in international trade are open account (OA), cash in advance (CIA), letter of credit (LOC), and documentary collection (DC) (Crozet et al., 2022). This chapter concentrates on LOC, a financial product issued by a bank that guarantees payment on behalf of the customer, thereby transferring all transaction risk to the bank (Mooney & Blodgett, 1995; Han et al., 2015).

12.2.3 Challenges in international trade

International trade encompasses a multitude of processes and participants, leading to a complex exchange of information and financial transactions (Heutger & Kückelhaus, 2018). This complexity remains, despite advancements in ICT, as many stages of the international trade information flow continue to depend on paper processes, which are inherently susceptible to errors and fraud, causing delays and losses (Ganne, 2018). A recent study by Cutlan et al. (2022) underscored that global commerce is still impeded by manual processes involving various documentation, necessitating repetitive communication between companies.

Even with the incorporation of ICT in customs clearance, there is still scope for enhancement, as the process involves multiple stages and documents and necessitates communication among multiple parties (Ganne, 2018). Indeed, Canham et al. (2022) discovered that 57% of traders concurred that cross-border processes added unnecessary and significant time to shipments, whilst 73% of traders agreed that within the next decade, customs processes would be finalised before shipments reach the border.

Trade finance, a crucial component of international trade, often incurs high costs due to its dependence on paper processes, making it labour-intensive (Ganne, 2018). A study by Ramachandran et al. (2017) revealed that a trade finance transaction involves more than 20 actors interacting with between 10–20 documents, resulting in approximately 5,000 interactions. Moreover, value-adding interactions accounted for only about 1%, whilst time-wasting interactions such as "ignore" or "transmit" constituted more than 85% of interactions.

Another factor contributing to the high costs and inefficiencies of trade finance processes is the number of verification checks required, as trust is nontransferable, and each party has different credit and reputation levels (Kowalski et al., 2021). Consequently, the costs and complexities have prompted various stakeholders to investigate the potential use of blockchain technology to possibly eliminate paperwork involved in the various processes of international trade (Ganne, 2018).

12.2.4 What is blockchain?

The term "blockchain" is used to denote a network managed by peer-to-peer nodes that utilise distributed ledger technology, where data and transactions are shared and secured using various cryptographic technologies (Ganne, 2018). Blockchains are safeguarded by various consensus algorithms, with the two most prevalent mechanisms being Proof-of-Work (PoW) and Proof-of-Stake (PoS) (Bach et al., 2018).

The PoW consensus mechanism necessitates the use of computing power via central processing units (CPUs) to generate (mine) new blocks on the network. New transactions are broadcast to all miners (nodes), who gather the transactions into a block before a miner is selected to work on generating a SHA-256 hash for each block. Consequently, the first node to hash the latest block broadcasts it to the network and adds it to the blockchain (Nakamoto, 2008).

Conversely, PoS consensus is a virtual process that involves a set of validators randomly taking turns to propose and vote on the next block, where the weight of each vote is dependent on the size of its stake (Buterin, 2017; Shah & Parveen, 2020).

12.2.5 Applications of blockchain

Blockchain technology has found applications in numerous sectors such as finance, governance, IoT, health, education, privacy and security, business and industry, and data management (Casino et al., 2019). The business and finance sectors include the international trade industry and thus, for the purpose of this chapter, the emphasis will be on the application of blockchain technology in the supply chain sector, and more specifically, in international trade, trade financing, and customs clearance.

The use of blockchain technology in the context of international trade is an emerging topic of interest in academic studies and shows potential to address the outstanding issues currently present in the international trade industry. Therefore, in the subsequent section, this chapter aims to address questions regarding the applications, benefits, and primary challenges of blockchain in the context of international trade.

12.3 APPLICATIONS, BENEFITS AND CHALLENGES

To delve into the applications, benefits, and challenges of blockchain in international trade, a comprehensive literature search was conducted using the SCOPUS database. The search incorporated keywords such as "supply chain", "international trade", "customs clearance", and "trade finance" in association with "blockchain". This procedure initially produced 1158 articles, which were subjected to rigorous screening and evaluation. Through this examination, the selection was narrowed down to 41 articles that substantially contributed to the content of this section.

12.3.1 Applications of blockchain technology in international trade

Blockchain technology, particularly its utilisation of smart contracts, enables the automation of various processes and could be employed to enhance international trade (Ganne, 2018). The applications of blockchain technology in international trade supply chains are well-researched – studies across multiple industries were discovered, such as textile and clothing (Agrawal et al., 2021), oil and gas (Ahamad et al., 2022; Aslam et al., 2021), food (Ali et al., 2021; Kumar et al., 2022), autonomous vehicles (Arunmozhi et al., 2022), maritime (Balci & Surucu-Balci, 2021), and luxury goods, among others. Literature also demonstrates that blockchain technology has been implemented in more specific processes such as customs clearance (Chang et al., 2020) and trade finance (Al Amaren et al., 2020; Bogucharskov et al., 2018).

Whilst academic research typically centres around proof-of-concept systems, literature covering industrial applications has demonstrated that blockchain technology has been successfully applied in supply chains at various scales, such as in the case of Walmart's food traceability project with Maersk and IBM and TradeLens' end-to-end trade document handling platform (Gonczol et al., 2020). Chang et al. (2020) discussed the utilisation of blockchain technology by T-Mining, a startup in Antwerp, for container release operations in the shipping industry. Accenture has also harnessed blockchain technology in supply chain networks to provide efficient document management (Chang et al., 2020).

The customs clearance process is fraught with paper-intensive procedures, leading to escalated costs and delays (Appeals & Struye de Swielande, 1998; Ganne, 2018). Research has been conducted around the application of blockchain technology in customs clearance processes, typically as part of a broader supply chain implementation (Al Amaren et al., 2020; Liu et al., 2021). For instance, Liu et al. (2022) carried out a study in which the customs clearance process of the Shanghai Yangshan Port was executed on the Hyperledger Fabric blockchain through smart contracts.

In the realm of trade finance, several studies have been undertaken on the feasibility of blockchain-based systems for the LOC process. For instance, Al Amaren et al. (2020) suggested the employment of blockchain technology for the issuance of LOC and automation through smart contracts. Toorajipour et al. (2022) put forth a comprehensive system that allows a buyer and seller to carry out business transactions entirely on the blockchain, eliminating the need for a third party for mediation or execution. Existing literature has also indicated that companies have been making use of blockchain technology for tasks such as managing trading documents (Chang et al., 2019b), Know Your Customer (KYC) and anti-money laundering (AML) checks, accounting and auditing, and real-time trade settlement (Rijanto, 2021).

12.3.2 Benefits of blockchain technology in international trade

Numerous benefits identified in the literature have been classified into seven key categories, outlined below.

12.3.2.1 Traceability and transparency

Public blockchains, such as Bitcoin, are inherently transparent, with all transactions being publicly declared and visible on the blockchain, thereby enabling anyone to scrutinise the origin and destination of transactions (Nakamoto, 2008). The traceability in supply chains is crucial as it enhances monitoring and visibility, assures the quality of products, eradicates fraud, fosters trust between business partners and consumers, and offers evidence of ethical sourcing and production (Razak et al., 2021).

The transparency of data on the blockchain also diminishes conflicts arising from disagreements and reduces administrative and documentation costs, as all parties have real-time access to the same data (Valeria et al., 2022). A wealth of research has demonstrated that blockchain technology can augment the traceability and transparency of supply chains across various industries (Agrawal et al., 2021; Ahamad et al., 2022; Balci & Surucu-Balci, 2021; Kumar et al., 2022), leading to enhanced supply chain visibility and resilience (Min, 2019). This characteristic also renders blockchain-based systems suitable for trade finance as it facilitates efficient and secure trade execution between buyers and sellers (Al Amaren et al., 2020).

12.3.2.2 Immutability

The immutability of blockchain signifies that once records are validated, they cannot be altered or removed, rendering it considerably more dependable than conventional database systems (Ganne, 2018). This characteristic renders trade and supply chains impervious to tampering, facilitates proper auditing and performance analysis (Ahamad et al., 2022; Rejeb et al., 2019), and diminishes the probability of fraud (Ali et al., 2021). Immutability also implies that smart contracts cannot be retrospectively altered or interfered with, thereby enabling regulators to utilise smart contracts to enforce regulatory compliance in supply chains (Chang et al., 2020). Gonczol et al. (2020) further expound that supply chain applications can reap benefits from the immutable property of blockchains due to the linkage between physical products and assets and their digital record on the blockchain.

12.3.2.3 Cost reduction

The employment of blockchain technology in global trade has been extensively documented to yield cost savings. For instance, the digitalisation of paper-heavy procedures prevalent in international trade results in enhanced efficiency and cost reductions in trade agreements and settlement (Al-Jaroodi & Mohamed, 2019; Chang et al., 2020), supply chains (Arunmozhi et al., 2022; Aslam et al., 2021; Kumar et al., 2020), trade finance (Bogucharskov et al., 2018; Chang et al., 2019b; Kowalski et al., 2021), and customs clearance (Chang et al., 2019a; Liu et al., 2021).

These cost savings typically arise due to faster process times and a reduction in administrative and paper-heavy tasks (Chang et al., 2020). Case studies conducted by Markus & Buijs (2022) discovered that the utilisation of smart contracts and blockchain technology led to a decrease in processes prone to errors and the disintermediation of various parties involved in data processing activities, thereby reducing operating costs.

12.3.2.4 Increased efficiency

Scholarly literature has demonstrated that the application of blockchain technology can enhance efficiency in diverse sectors, such as construction and telecommunication (Al-Jaroodi & Mohamed, 2019), trade finance (Al Amaren et al., 2020), and supply chains, shipping, and logistics (Balci & Surucu-Balci, 2021; Deng & Ouyang, 2022), among others. The digitisation of paper-intensive tasks through the use of smart contracts, which facilitate the automation of processes and contracts, consequently, reduces labour costs and augments process efficiency (Valeria et al., 2022). An instance of this is CargoChain, a software that digitises supply chain documentation and automates the typical document handling processes, resulting in brief reaction times and enhanced work throughput (Tönnissen & Teuteberg, 2020). Liu et al. (2022) discovered that by implementing the customs clearance process of the Shanghai Yangshan Port on the Hyperledger Fabric blockchain via smart contracts, the time required for document validation was reduced from 2–3 days to merely a few hours.

12.3.2.5 Reduction of human errors

A prevalent theme identified in scholarly works is that the employment of blockchain technology and smart contracts leads to a decrease in human errors. Shared distributed ledgers facilitate straightforward verification, storage, and auditing of supply chain data, thereby diminishing the potential for human errors that could escalate costs and cause delays (Al-Jaroodi & Mohamed, 2019; Min, 2019).

In the realm of trade finance, where human involvement is typically crucial to ensure accuracy, Kowalski et al. (2021) suggested the use of blockchain technology in tandem with IoT to minimise human input and thereby reduce the scope for errors. Central banks such as the Monetary Authority of Singapore (MAS) and the Hong Kong Monetary Authority (HKMA) have been conducting research and trials on the application of distributed ledger technology in trade finance with the aim of reducing human error and fraud (Kumar et al., 2020).

12.3.3 Challenges of blockchain technology in international trade

Whilst blockchain technology provides a multitude of advantages, there remain challenges that require attention, as elaborated in the ensuing discussion.

12.3.3.1 High cost of investment

Firms encounter the paradox of blockchain technology diminishing operational and administrative expenses, yet with a substantial initial investment cost (Sunny et al., 2020; Wang et al., 2019). A case study by Ali et al. (2021) discovered that whilst SMEs in the food supply chain industry could reap benefits from blockchain technology, it might lead to narrow profit margins due to the considerable investment needed for its implementation. Kumar et al. (2020) further deduced that the cost-effectiveness of implementing blockchain technology for international trade ought to be scrutinised on an individual basis, given the high overhead costs. A decision-making trial and evaluation laboratory analysis, utilising "matrixes or diagrams…" as a "structural modelling approach" (Si et al., 2018), determined that the elevated investment cost of blockchain implementation is a consequence of development, training, and hardware and software acquisitions to bolster the new infrastructure (Kumar et al., 2022).

12.3.3.2 Lack of regulation

Whilst blockchain technology can be utilised to aid in regulation compliance (Ahmad et al., 2022; Al-Jaroodi & Mohamed, 2019; Chang et al., 2020), the regulations and legislations surrounding the use of blockchain technology and the legal validity of transactions and smart contracts remain ambiguous (Ganne, 2018). This results in a lack of trust by organisations contemplating the implementation of blockchain technology until adequate government regulations are put into place (Balci & Surucu-Balci, 2021).

An additional challenge is the unclear delineation of jurisdiction for each transaction on the blockchain network, given that each node can exist in various geolocations (Chang et al., 2020). A survey conducted by Petersen et al. (2018) of participants in the supply chain sector revealed that 56% concur that "regulatory uncertainty" is a hindrance to blockchain adoption. Thus, regulatory frameworks could offer further clarity in this area.

12.3.3.3 Scalability

The scalability of blockchains has been identified as a barrier and challenge to the adoption of blockchain technology, due to problems with data handling and the low throughput of public blockchains (Al-Jaroodi & Mohamed, 2019; Chang et al., 2020; Liu et al., 2021). A study conducted by Kumar et al. (2022) discovered that the "lack of scalability" was the "third most significant barrier" to blockchain adoption in supply chains. However, this has been demonstrated to be less of an issue with private blockchain infrastructure with optimised hardware and node distribution (Ganne, 2018). In fact, newer blockchain architectures have exhibited an increase in throughput and a reduction in latency (Gonczol et al., 2020).

12.3.3.4 Sustainability

In scholarly works, the sustainability of blockchain technology has been scrutinised, both in terms of its longevity in industrial applications and the quantity of energy required to sustain blockchain networks (Ghode et al., 2020; Kowalski et al., 2021; Sunny et al., 2020). Regarding high energy consumption, such assertions are based on the energy usage of public blockchains like Bitcoin, which employ a PoW consensus architecture. More recent blockchains utilise PoS consensus, which necessitates significantly less energy (Ganne, 2018) and is indeed a focal point of recent advancements in blockchain technology (Valle & Oliver, 2021).

12.3.3.5 Interoperability

Interoperability, particularly in international trade and supply chains, is crucial due to the involvement of various stakeholders and the presence of data on diverse platforms and ledgers. Blockchains are further divided and inherently incapable of intercommunication, a result of efforts to address blockchain scalability and the requirements of different institutions (Ahmad et al., 2022; Chang et al., 2020; Ganne, 2018). Another obstacle is the interoperability between blockchains and traditional data management systems, which renders the integration process intricate and expensive (Al-Jaroodi & Mohamed, 2019). Nonetheless, recent advancements and research in the realm of blockchain interoperability have been promising, with public blockchain ecosystems such as Cosmos and Polkadot pioneering cross-chain communication (Ahmad et al., 2022; Chang et al., 2020).

12.3.3.6 Security and privacy

In spite of its inherent robustness when compared to conventional database systems (Ganne, 2018), blockchain technology still poses certain security and privacy risks. A majority of previous research identifies the risk associated with smart contracts as the primary security threat to blockchains, as they are executed through programming languages and are therefore susceptible to external cyber-attacks, bugs, and errors (Ahmad et al., 2022; Al-Jaroodi & Mohamed, 2019; Kowalski et al., 2021; Liu et al., 2021). Other security risks, such as 51% attacks on a blockchain network, are significantly higher on smaller public networks with a smaller set of nodes but can be alleviated using privately managed permissioned blockchains (Al-Jaroodi & Mohamed, 2019; Ganne, 2018). The inherent characteristics of blockchain technology, such as traceability and transparency, could potentially result in data privacy issues, thus necessitating stakeholders involved to ensure the security of such data (Al-Jaroodi & Mohamed, 2019; Liu et al., 2021). This has also been demonstrated to be a hindrance in the adoption of blockchain due to the preference of certain organisations to maintain control and protection of information (Ghode et al., 2020).

12.3.3.7 Complexity

The adoption of blockchain technology for global trade has encountered opposition, owing to the industry's fragmented nature, which involves numerous stakeholders, documentation processes, and implementation complexities (Ali et al., 2021; Chang et al., 2020; Liu et al., 2021). A study conducted by Balci & Surucu-Balci (2021) revealed that the majority shareholders' lack of understanding about blockchain technology and its advantages presents a significant obstacle to the technology's adoption. This is further supported by a separate interview of supply chain experts by Wang et al. (2019), who discovered a general lack of knowledge concerning "blockchain technicalities, functions, or benefits". Another survey by Kumar et al. (2022) identified the "low competency of workers" as a significant barrier to the adoption of blockchain in food supply chains. Part of the complexity of blockchain technology is ascribed to the concept of private key management, as the loss of private keys leads to the complete loss of any data associated with those keys (Liu et al., 2021).

12.3.4 Summary of findings and implications for management and theory

The details provided in this section are encapsulated in Figure 12.1, which depicts the utilisation of blockchain technology in diverse aspects of global trade, such as supply chains, trade finance, and customs clearance, together with its corresponding advantages and difficulties. The discussed advantages of blockchain technology in international trade include traceability and transparency, immutability, cost reduction, enhanced efficiency, and the minimisation of human errors. Conversely, the challenges obstructing the implementation of blockchain were also deliberated and comprise its prohibitive investment cost, absence of regulation, scalability, sustainability, interoperability, security, privacy, and complexity issues.

Blockchain's application in global trade operations presents numerous managerial implications. It influences strategic planning through the understanding of its potential benefits and challenges as presented in Figure 12.1. The significant initial investment cost calls for careful financial planning and risk assessment on the part of the management and the regulatory void around this technology highlights the need for staying up to date with potential legislative changes.

As blockchain technology promises improved efficiency and error reduction, potentially strengthening operational processes, it also raises security and privacy concerns, necessitating prioritised measures to protect sensitive data. From the managerial perspective, addressing scalability and sustainability issues is crucial for the long-term viability of blockchain-based systems. The challenge of interoperability with existing systems is also a consideration when integrating this technology.

Findings from this chapter have a potential to enhance the technological adoption theory by providing insights into blockchain technology adoption factors. Academics may find value in the framework presented in Figure 12.1 as it explains blockchain's role in international trade. Potential further research avenues exist for exploring blockchain's traceability within the context of information systems theory. Meanwhile, the absence of regulation may present a fruitful study area for policy theory.

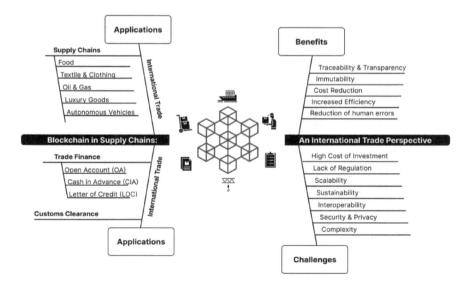

Figure 12.1 Blockchain in international trade: applications, benefits, and challenges.

Exploring sustainability issues around use of blockchain could contribute to theory development in sustainable business practices. Finally, addressing the security and privacy concerns could intersect fields of computer science and technology implementation.

To further illustrate these points, the following section presents a case study, providing a practical exploration of these managerial and theoretical considerations and their real-world implications.

12.4 CASE STUDY: AZARC'S BLOCKCHAIN SOLUTION AND ECOSYSTEM OF TRUST

In recent years, the international trade landscape worldwide has experienced a significant shift towards digitalisation and automation. Azarc,[1] a firm specialising in business process automation with a focus on streamlining cross-border supply chains through digital automation, plays a role in this revolution. This case study explores Azarc's innovative application of blockchain technology, specifically its Verathread middleware, to tackle the challenges posed by the UK's post-Brexit manual customs clearance processes.

Operating in a sector where efficiency, speed, and cost reduction are of utmost importance, Azarc's Verathread middleware, a decentralised workflow solution that utilises blockchain and operates the Rune Utility Trade Network, has become a crucial part of the company's offerings. It connects systems and businesses seamlessly, enabling autonomous data transactions.

Following United Kingdom's departure from the European Union, a new UK's 2025 Border Strategy[2] aims to create a world-leading border by harnessing data and technology to facilitate trade whilst ensuring biosecurity and security. Azarc actively engaged in testing a new border model, the Ecosystem of Trust, designed to minimise trade frictions across public and private sectors.

During the testing phase, Azarc concentrated on two vital trade routes: the European Union and the Rest-of-the-World, operating via the ports of Harwich and London Gateway. Fresh fruits and vegetables, along with charcoal and wood, were selected as the commodities for this study to represent a diverse array of products with different clearance requirements.

Azarc deployed its Verathread solution to facilitate direct data contribution from supply chain actors, enhancing clearance efficiency by providing increased pre-clearance data and real-time visibility of goods movement across borders. The system flawlessly captured data from various sources, including IoT devices, creating a comprehensive and transparent record of the entire supply chain.

The integration of channel partners using blockchain technology through Verathread yielded several benefits for Azarc and the overall trade ecosystem. Primarily, the real-time data contribution from supply chain actors significantly improved clearance efficiency. This increased pre-clearance data not only expedited the clearance process but also enhanced the accuracy of customs declarations.

Moreover, the decentralised nature of the blockchain solution eliminated the need for upgrades to legacy systems, reducing the burden on businesses and ensuring a smooth transition to the new system. The removal of friction at the border, coupled with the ability to facilitate real-time payment of duties and taxes, resulted in cost savings for traders.

Azarc's successful implementation of blockchain technology in the international trade domain, particularly in customs clearance processes, demonstrates the transformative potential of decentralised workflows. By actively participating in the Ecosystem of Trust and leveraging the Verathread, Azarc not only addressed the challenges posed by the manual customs clearance mess but also contributed to the realisation of the UK's 2025 Border Strategy and development of regulatory frameworks surrounding use of blockchain solutions in customs clearance. This case study underscores the importance of innovative technologies, such as blockchain, in creating more efficient, transparent, and cost-effective international trade ecosystems.

12.5 CONCLUSIONS

This chapter examined utilisation of blockchain technology within the sphere of international trade, delving into its applications, advantages, and obstacles. Blockchain has proven its worth across various sectors of international trade, including food, textiles, oil and gas, luxury items, and supply chains for autonomous vehicles. Importantly, it has been effectively used in facilitating LOC transactions in trade finance and procedures for customs clearance. These applications lead to benefits such as improved traceability, transparency in transactions, security, prevention of fraud, increased efficiency, and a decreased dependence on processes that require extensive paperwork, resulting in cost savings. Despite these benefits, there are still challenges in the implementation of blockchain technology in international trade, which include substantial investment costs, regulatory gaps, scalability issues, sustainability concerns, interoperability challenges, issues with data security and privacy, and the complexity of the technology. Whilst these challenges pose barriers, ongoing advancements, as illustrated by the Azarc case study, display successful applications of blockchain in this field.

NOTES

1. AZARC: https://www.azarc.io/.
2. 2025 UK Border Strategy: https://www.gov.uk/government/publications/2025-uk-border-strategy.

REFERENCES

Agrawal, T. K., Kumar, V., Pal, R., Wang, L., & Chen, Y. (2021). Blockchain-based framework for supply chain traceability: A case example of textile and clothing industry. *Computers & Industrial Engineering*, 154, 107130. https://doi.org/10.1016/j.cie.2021.107130.

Ahamad, R., Salah, K., Jayaraman, R., Yaqoob, I., & Omar, M. (2022). Blockchain in oil and gas industry: Applications, challenges, and future trends. *Technology in Society*, 101941.

Ahmad, R. W., Salah, K., Jayaraman, R., Yaqoob, I., & Omar, M. (2022). Blockchain in oil and gas industry: Applications, challenges, and future trends. *Technology in Society*, 68, 101941. https://doi.org/10.1016/j.techsoc.2022.101941

Al Amaren, E., Md Ismail, C. T., & Md Nor, M. Z. (2020). The blockchain revolution: A game-changing in letter of credit (L/C)? *International Journal of Scientific & Technology Research*, 29, 6052–6058.

Ali, M. H., Chung, L., Kumar, A., Zailani, S., & Tan, K. H. (2021). A sustainable blockchain framework for the halal food supply chain: Lessons from Malaysia. *Technological Forecasting and Social Change*, 170, 120870.

Al-Jaroodi, J., & Mohamed, N. (2019). Blockchain in industries: A survey. *IEEE Access*, 7, 36500–36515. https://doi.org/10.1109/ACCESS.2019.2903554

Appeals, T., & Struye de Swielande, H. (1998). Rolling back the frontiers: The customs clearance revolution. *The International Journal of Logistics Management*, 9, 111–118.

Arunmozhi, M., Venkatesh, V. G., Arisian, S., Shi, Y., & Raja Sreedharan, V. (2022). Application of blockchain and smart contracts in autonomous vehicle supply chains: An experimental design. *Transportation Research Part E: Logistics and Transportation Review*, 165, 102864.

Aslam, J., Saleem, A., Khan, N. T., & Kim, Y. B. (2021). Factors influencing blockchain adoption in supply chain management practices: A study based on the oil industry. *Journal of Innovation & Knowledge*, 6(2), 124–134. https://doi.org/10.1016/j.jik.2021.01.002

Bach, L. M., Mihaljevic, B., & Zagar, M. (2018). Comparative analysis of blockchain consensus algorithms. *2018 41st International Convention on Information and Communication Technology, Electronics and Microelectronics (MIPRO)*. Opatija, Croatia.

Balci, G., & Surucu-Balci, E. (2021). Blockchain adoption in the maritime supply chain: Examining barriers and salient stakeholders in containerized international trade. *Transportation Research Part E: Logistics and Transportation Review*, 156, 102539. https://doi.org/10.1016/j.tre.2021.102539

Bogucharskov, A. V., Pokamestov, I. E., Adamova, K. R., & Tropina, Zh. N. (2018). Adoption of blockchain technology in trade finance process. *Journal of Reviews on Global Economics*, 7, 510–515.

Boschian, V., Fanti, M. P., Iacobellis, G., & Ukovich, W. (2010). The assessment of ICT solutions in customs clearance operations. *2010 IEEE International Conference on Systems, Man and Cybernetics*. Istanbul, Turkey.

Buterin, V. (2017). Proof of Stake FAQ. https://web.archive.org/web/20220904234336/https://vitalik.ca/general/2017/12/31/pos_faq.html#what-is-proof-of-stake

Canham, J., Ellanti, P., & Fanguy, M. (2022). *Borders 2030: From vision to reality.* Accenture.

Casino, F., Dasaklis, T. K., & Patsakis, C. (2019). A systematic literature review of blockchain-based applications: Current status, classification and open issues. *Telematics and Informatics*, 36, 55–81. https://doi.org/10.1016/j.tele.2018.11.006

Chang, S. E., Chen, Y.-C., & Lu, M.-F. (2019a). Supply chain re-engineering using blockchain technology: A case of smart contract based tracking process. *Technological Forecasting and Social Change*, 144, 1–11. https://doi.org/10.1016/j.techfore.2019.03.015

Chang, S. E., Luo, H. L., & Chen, Y. (2019b). Blockchain-enabled trade finance innovation: A potential paradigm shift on using letter of credit. *Sustainability*, 12(1), 188. https://doi.org/10.3390/su12010188

Chang, Y., Iakovou, E., & Shi, W. (2020). Blockchain in global supply chains and cross border trade: A critical synthesis of the state-of-the-art, challenges and opportunities. *International Journal of Production Research*, 58(7), 2082–2099. https://doi.org/10.1080/00207543.2019.1651946

Committee on the Global Financial System. (2014). *Trade Finance: Developments and Issues.* Bank for International Settlements.

Crozet, M., Demir, B., & Javorcik, B. (2022). International trade and letters of credit: A double-edged sword in times of crises. *IMF Economic Review*, 70, 185–211.

Cutlan, M., Sanghvi, P., & Fahey, T. (2022). *At one: One connected supply chain. One big move forward.* Accenture.

Deng, X., & Ouyang, Y. (2022). Cross-border supply chain system constructed by complex computer blockchain for international cooperation. *Computational Intelligence and Neuroscience*, 2022, 1–10. https://doi.org/10.1155/2022/6221211

Dodzin, S., & Vamvakidis, A. (2004). Trade and industrialization in developing economies. *Journal of Development Economics*, 75, 319–328.

Gani, A. (2015). *The Efficiency of Customs Clearance Processes Can Matter for Trade.* International Atlantic Economic Society.

Ganne, E. (2018). *Can Blockchain Revolutionize International Trade?* WTO Publications.

Gereffi, G. (1999). International trade and industrial upgrading in the apparel commodity chain. *Journal of International Economies*, 48, 37–70.

Ghode, D. J., Yadav, V., Jain, R., & Soni, G. (2020). Blockchain adoption in the supply chain: An appraisal on challenges. *Journal of Manufacturing Technology Management*, 32(1), 42–62. https://doi.org/10.1108/JMTM-11-2019-0395

Gonczol, P., Katsikouli, P., Herskind, L., & Dragoni, N. (2020). Blockchain implementations and use cases for supply chains-A survey. *IEEE Access*, 8, 11856–11871. https://doi.org/10.1109/ACCESS.2020.2964880

Han, C. R., Nelen, H., & Joo, M. Y. (2015). Documentary credit fraud against banks: analysis of Korean cases. *Journal of Money Laundering Control*, 18(4), 457–474.

Heutger, M., & Kückelhaus, M. (2018). *Blockchain in Logistics.* DHL Customer Solutions & Innovation.

Irwin, D. A., & Terviö, M. (2002). Does trade raise income? *Journal of International Economics*, 58, 1–18.

Kowalski, M., Lee, Z. W. Y., & Chan, T. K. H. (2021). Blockchain technology and trust relationships in trade finance. *Technological Forecasting and Social Change*, 166, 120641. https://doi.org/10.1016/j.techfore.2021.120641

Kumar, A., Liu, R., & Shan, Z. (2020). Is blockchain a silver bullet for supply chain management? Technical challenges and research opportunities. *Decision Sciences*, 51(1), 8–37. https://doi.org/10.1111/deci.12396

Kumar, S., Raut, R. D., Agrawal, N., Cheikhrouhou, N., Sharma, M., & Daim, T. (2022). Integrated blockchain and internet of things in the food supply chain: Adoption barriers. *Technovation*, 118, 102589. https://doi.org/10.1016/j.technovation.2022.102589

Latif, Z., Mengke, Y., Latif, S., Ximei, L., Pathan, Z. H., Salam, S., & Jianqiu, Z. (2018). The dynamics of ICT, foreign direct investment, globalization and economic growth: Panel estimation robust to heterogeneity and cross-sectional dependence. *Telematics and Informatics*, 35, 318–328.

Li, G., & Li, N. (2019). Customs classification for cross-border e-commerce based on text-image adaptive convolutional neural network. *Electronic Commerce Research*, 19, 779–800.

Liu, J., Zhang, H., & Zhen, L. (2021). Blockchain technology in maritime supply chains: Applications, architecture and challenges. *International Journal of Production Research*, 1–17. https://doi.org/10.1080/00207543.2021.1930239

Liu, Y., Zhou, Z., Yang, Y., & Ma, Y. (2022). Verifying the smart contracts of the port supply chain system based on probabilistic model checking. *Systems*, 10(1), 19. https://doi.org/10.3390/systems10010019

Markus, S., & Buijs, P. (2022). Beyond the hype: How blockchain affects supply chain performance. *Supply Chain Management: An International Journal*, 27(7), 177–193. https://doi.org/10.1108/SCM-03-2022-0109

Min, H. (2019). Blockchain technology for enhancing supply chain resilience. *Business Horizons*, 62, 35–45.

Mooney, J. L., & Blodgett, M. S. (1995). Letters of credit in the global economy: Implications for international trade. *Journal of International Accounting, Auditing and Taxation*, 4, 175183.

Nakamoto, S. (2008). Bitcoin: A peer-to-peer electronic cash system. Technical Report. [Online]. Available: https://bitcoin.org/bitcoin.pdf

Norberg, H. C. (2019). *Unblocking The Bottlenecks And Making The Global Supply Chain Transparent: How Blockchain Technology Can Update Global Trade*. University of Calgary.

Okazaki, Y. (2018). *Unveiling the Potential of Blockchain for Customs*. WCO.

Petersen, M., Hackius, N., & von See, B. (2018). Mapping the sea of opportunities: Blockchain in supply chain and logistics. *It - Information Technology*, 60(5–6), 263–271. https://doi.org/10.1515/itit-2017-0031

Ramachandran, S., Porter, J., Kort, R., Hanspal, R., & Garg, H. (2017). *Digital Innovation in Trade Finance*. Boston Consulting Group.

Razak, G. M., Hendry, L. C., & Stevenson, M. (2021). Supply chain traceability: A review of the benefits and its relationship with supply chain resilience. *Production Planning and Control. Scopus.* https://doi.org/10.1080/09537287.2021.1983661

Rejeb, A., Keogh, J. G., & Treiblmaier, H. (2019). Leveraging the internet of things and blockchain technology in supply chain management. *Future Internet*, 11(7), 161. https://doi.org/10.3390/fi11070161

Rijanto, A. (2021). Blockchain technology adoption in supply chain finance. *Journal of Theoretical and Applied Electronic Commerce Research*, 16(7), 3078–3098. https://doi.org/10.3390/jtaer16070168

Robinson, R., Allais, M., Bertrand, T., Balassa, B., & Wonnacott, B. (2020). International Trade. In *Encyclopaedia Britannica*. https://www.britannica.com/money/international-trade

Saberi, S., Kouhizadeh, M., Sarkis, J., & Shen, L. (2019). Blockchain technology and its relationships to sustainable supply chain management. *International Journal of Production Research*, 57(7), 2117–2135. https://doi.org/10.1080/00207543.2018.1533261

Saheb, T., & Mamaghani, F. H. (2021). Exploring the barriers and organizational values of blockchain adoption in the banking industry. *The Journal of High Technology Management Research*, 32, 100417.

Samsir, N. A., & Jain, A. A. (2023). Overview of blockchain technology and its application in healthcare sector. In: Joshi, A., Mahmud, M., Ragel, R.G. (Eds.) *Information and Communication Technology for Competitive Strategies (ICTCS 2021). Lecture Notes in Networks and Systems*, vol 400. Springer, Singapore. https://doi.org/10.1007/978-981-19-0095-2_63, pp. 661–668.

Serena, J. M., & Vasishtha, G. (2015). What drives bank-intermediated trade finance? Evidence from cross-country analysis (September 16, 2015). Banco de Espana Working Paper No. 1524, Available at SSRN: https://ssrn.com/abstract=2661370 or http://dx.doi.org/10.2139/ssrn.2661370. *SSRN Electronic Journal*.

Shah, J., & Parveen, S. (2020). Understanding the blockchain technology beyond bitcoin. In R. K. Phanden, K. Mathiyazhagan, & R. Kumar (Eds.), *Advances in Industrial and Production Engineering* (pp. 499–516). Springer.

Si, S.-L., You, X.-Y., Liu, H.-C., & Zhang, P. (2018). DEMATEL technique: A systematic review of the state-of-the-art literature on methodologies and applications. *Mathematical Problems in Engineering*, 2018, e3696457. https://doi.org/10.1155/2018/3696457

Statista. (2020). *International Seaborne Trade Carried by Container Ships from 1980 to 2020 (in billion tons loaded)*. Statista.

Sunny, J., Undralla, N., & Madhusudanan Pillai, V. (2020). Supply chain transparency through blockchain-based traceability: An overview with demonstration. *Computers & Industrial Engineering*, 150, 106895. https://doi.org/10.1016/j.cie.2020.106895

Tönnissen, S., & Teuteberg, F. (2020). Analysing the impact of blockchain-technology for operations and supply chain management: An explanatory model drawn from multiple case studies. *International Journal of Information Management*, 52, 101953. https://doi.org/10.1016/j.ijinfomgt.2019.05.009

Toorajipour, R., Oghazi, P., Sohrabpour, V., Patel, P. C., & Mostaghel, R. (2022). Block by block: A blockchain-based peer-to-peer business transaction for international trade. *Technological Forecasting and Social Change*, 180. Scopus. https://doi.org/10.1016/j.techfore.2022.121714

USSC. (2018). Bitcoin Glossary. https://www.ussc.gov/sites/default/files/pdf/training/annual-national-training-seminar/2018/Emerging_Tech_Bitcoin_Crypto.pdf

Valeria, S., Vitaliia, D., Kateryna, T., Rostyslav, H., & Oleg, Y. (2022). The impact of blockchain technology on international trade and financial business. *Universal Journal of Accounting and Finance*, 10(1), 102–112. https://www.hrpub.org/journals/article_info.php?aid=11672

Valle, F. D., & Oliver, M. (2021). Blockchain-based information management for supply chain data-platforms. *Applied Sciences (Switzerland)*, 11(17). Scopus. https://doi.org/10.3390/app11178161

Wang, Y., Singgih, M., Wang, J., & Rit, M. (2019). Making sense of blockchain technology: How will it transform supply chains? *International Journal of Production Economics*, 211, 221–236. https://doi.org/10.1016/j.ijpe.2019.02.002

Index

Note: **Bold** page numbers refer to tables and *italic* ones to figures.

academics/academy 251
　journals 220
　literature 234
　penetration 221–222
　theoretical implications for 23
accountability 68, 73, 91, 94–95, 182, 191, 209, 240
accreditation 183
Adaptive Boosting (AdaBoost) 95, 96, 99
agricultural/agriculture 178, 180
　landscape of 181
　products 193–194
　supply chains 178, 185, 194
agri-food 193
　industry 179–180
　landscape 194
　supply chain 179
Alzheimer's disease 105
animal
　health 181
　management 182
anti-money laundering (AML) 70, 130, 245
applicability of blockchain 193–194
artificial intelligence 77, 142, 158, 223; *see also* healthcare sector, artificial intelligence and big data application in
　applications of 105–106
　technologies in vaccine supply chain management 90
　utilizing 118
asset-reference tokens 70

asymmetric information 179
augmented reality 186, 190–191
authentication 6, 25, 36, 178
automation 209, 240, 245, 247, 252
autonomy 199

bandwidth 161–162
bank/banking 5, 37, 70, 154, 159, 210–211, 240, 243, 247
bibliographic collection 219–220, 224–225
bibliometric analysis 4, 10–11, **10**, *12*, 219–221
　procedure *12*
　on supply chain 9
Bibliometrix 17, *221*, *222*, *224*
'Biblioshiny' application 220, 224
big data and artificial intelligence (AI) 201, *see also* healthcare sector, artificial intelligence and big data application in
　academic literature 106–107
　analytics 118
　challenges in implementing 106, 114
　conceptualization of 117
　convergence of 123
　description of 105
　effective deployment of 123
　framework for 122, *122*
　structures 119
　studies on convergence of 107, **108–113**
　systematic review of convergence of 106–107, *107*

259

Index

biometrics 115
Bitcoin 3, 50, 66–69, 151, 154–155, 159–162, 179, 246, 249; *see also* cryptocurrency
  adoption of 155
  advent of 154
  concept of 49
  decentralised nature of 70
  emergence of 154
  inception of 240
  privacy and security issues of 9
  source code for 68
blockchain
  advantages of 160
  applications of 219, 234, 244, *251*
  components of 8
  definition of 244
  disadvantages of 160
  enablers of 73, **73**
  features 202
  immutable property of 246
  implementation framework 164, *165*
  knowledge 163
  ledgers 132
  performance, measurement 153
  platform 71, 91, 94, 158, 160, 167, 170, 171
  readiness index 164
  structure *35*
Blockchain 1.0 154–155, **157**
Blockchain 2.0 155–156, **157**
Blockchain 3.0 156–157, **157**
Blockchain 4.0 157, **157**
blockchain adoption
  cryptocurrency as digital asset 67–69
  decision-making in 164
  description of 66–67
  digital assets, models for 74–76
  innovators and early adopters 69–70
  readiness index 164
  in supply chain 70–73
  technological–organization–environmental 163
Blockchain-as-a-service (BaaS) 158
blockchain enablers
  cryptocurrency adoption, strategy and managerial aspects of 77–78
  facilitation of 77
  macro drivers of 77
  organisational readiness and cryptocurrency 79–80
blockchain implementation framework
  academic contribution 169
  Blockchain 1.0 154–155
  Blockchain 2.0 155–156
  Blockchain 3.0 156–157
  Blockchain 4.0 157
  description of 151–153, 164
  evaluation 168
  evolution of 153–154
  ideation 164–166
  implementation 167–168
  improvement planning 169
  optimization 168–169
  practical contribution 169
  strategy 166–167
  technology selection 167
blockchain readiness model 40–45
blockchain technology (BCT) 34–35, 49, 152–153, 183, 200, 204–205, 234
  adoption 38–40
  advantages of 194
  articles as per year of publishing *206*
  benefits of 35–36
  categories 10, **10**
  characteristics of 9, 218–219
  community involvement in 164
  components of 9
  concept of 198
  consortium 159
  contributions of study 52
  core metrics *see* core metrics
  corporate investment in 3
  deep knowledge of 53
  demerits of 53
  description of 49–51
  disruptive and popular nature of 153
  drawbacks of 194
  future directions 62
  implementation of *203*, *205*, 253
  incorporation of 241
  industrial applications of 36–38
  integration of 50
  interruptions 50–53, **54–55**, 56–59, **57**, 61–62
  limitations 61–62
  literature 52–53, 153

managerial and theoretical
    implications 60
methodology 53–57
network security 58–59
performance of 153
potential use of 243
private 159
process of implementing 164–165
public 159
publications in 13, 15, **15**, **16**
research 51–52
results and discussion 57–60
types of 158
utilisation of 250
word cloud with blockchain
    features *205*
blockchain technology studies
    background and literature 4–5
    bibliometric analysis 10–11, **10**, *12*
    categories 10, *11*
    characteristics 5–9
    co-citation network *see* co-citation
        network
    cooperation network, mapping of
        13, *14*
    countries with contributions in
        14, **14**
    description of 3–4, 18–23
    document in 12, *13*
    highly cited publications in 15, **16**
    keywords 17, *17*, 20, **21**
    managerial and theoretical
        implications 23–24
    material and method 9–10
    publications in 12, *13*, 15, **15**
    results 11–18
    scientometric analysis 10–11, **10**, *12*
    sources in 22, *23*
    three-field plot 18, *19*
brainstorming 166
bugs 250
business 35, 69–73, 156, 160, 166–167,
    169–171, 185–186, 209–210
    corporations 210
    strategy 80, 167
    transparency 186

carbon emission 200, 203, 206
    data 38
    of greenhouse gasses 200
carbon footprints 38, 199–200, 203,
    207–208

carcasses, handling of 182
CargoChain 247
cash in advance (CIA) 243
Centers for Disease Control and
    Prevention (CDC) 89
central authority 5, 130, 159, 208
central processing units (CPUs) 244
channel partners 232, 252–253
chronic diseases 229–230
claims 114, 120, 227, 231–233
Clarivate Analytics 10
clearance efficiency 252–253
climate emergency 198, 200, 203,
    206–207
climatic emissions 217
co-authorship 13, 220
co-citation network
    of authors 17, *20*
    of references 18, *18*
    of sources 15, *19*
Codex Alimentarius Commission 181
collaboration 36, 38, 40, 44, 119
    data sharing 120
    interoperability 120
    openness to 71
    smart contracts 120
concatenation 4–9, 25
confidentiality 42, 151, 209
consensus protocol 9, 156, 158,
    163, **163**
consignment 188–190
consortium blockchain 159, 227, 240
construction 52, 133, 247
consumer 190–191, 207
    benefit 210
    feedback 186
consumption 178
    energy 155, 161, 200–202
    resource 161
    time 152
contagious diseases 229
contamination 182–183, 233
content analysis 225
continuous improvement 171
contract/contractual
    for difference 155
    farming 184
    obligations 184
conventional technology adoption
    frameworks 171
co-occurrence network 222–223,
    222, 225

core metrics 160
  fault tolerance 163, **163**
  finality 161
  latency 160–161
  resource consumption 161
  scalability 162
  throughput 160
corruption in healthcare 225–227, 230–231
cost reduction/savings 76, 153, 184, 246–247, 250–253
counterfeit drugs 121, 226–227, 230–231, 233
country scientific production 223, *223*
COVID-19 90, 231
  crisis 118
  pandemic 116, 122–123
  vaccine distribution data 90–91
crypto assets, development of 70
cryptocurrency 36, 50, 66–68, 70, 78, 154, 159, 179–180, 198
  adoption 68–69, 77–79, **79**
  classification of 67
  community involvement in 164
  for financial contracts 155
  implementation 78
  prices 69
  volatility 80
cryptographic/cryptography 71
  algorithm 5–6
  assets 68
  chain 49, 231
  coins 77
  currencies 68
  hash functions 7–8, 132, 218
  protocols 201–202
customers 77, 178
  education 80
  empowerment 229–230
customs clearance 242–243, 245, 247
  blockchain solutions in 253
  declarations 253
  efficiency of 242
  procedure 241
cyber-attacks 42, 250

DAG-based systems 158
data analysis, healthcare 220
  academic penetration 221–222
  co-occurrence network 222–223, *222*

country scientific production 223, *223*
descriptive analysis 220–221, *221*
literature on 220, *221*
word cloud analysis 224–225, *224*
data/databases
  acquisition 201
  authenticity and accuracy 229
  carbon emissions 38
  collection, process of 40
  decentralization of 38, 218
  energy consumption 201
  integrity 3, 73
  interoperability 218
  management 218
  manipulation 118–119
  mining 201
  monetization 117–118
  organized and objective 230
  ownership 117
  privacy 119, 232, 250
  protection 37
  relational 179
  retrieval and sharing 226
  rights protection 117
  security 35–37, 115, 151, 232
  sharing 120, 206–207
  storage and validation 208
  unstructured 219
decentralised applications (DApps) 135, 155, 210
decentralised automation energy (DAE) community 199
decentralized/decentralization 5, 72–73, 162, 199, 201, 206–208
  applications 156
  blockchain frameworks 121
  consensus algorithms 159
  of data and transactions 33–35
  energy grids 208
  finance 198
  functional attributes of 153
  storage 218
deep convolutional neural network with pool layer 97
deep learning (DL) 118
  algorithms 122
  models 93
deep neural network 95, 97
delegated proof-of-stake (DPoS) 163, 199
Delphi technique 166

Index

demand
  response programs 199
  variability 89
descriptive analysis 220–221, *221*
Design Science Research (DSR) framework 133
design thinking 166
Dhonor Healthtech 121
Diffusion of Innovations (DOI) framework 167–168
digital/digitization/digitalization 49, 252
  assets 66–67
  currency 67–68, 179
  of health records 224
  identities 9, 37, 209
  literacy 234
  penetration 234
  record-keeping 115–116
  signatures 5–9
  storage 224
  systems 241
  technology 50, 61, 77, 133
  tokens 68, 209
  transactions 151, 198, 240
  transformation management 169
  trust 115
digital supply chain (DSC) 53
digital technologies 50, 61, 77, 133, 224
digital transformation 37, 38, 67, 79, 90, 91, 120, 151, 169, 198, 240
Directional Acrylic Graph (DAG) 158
disaster management 210
disruption 35, 93
distributed application (DApp) platform 199
distributed ledger technology (DLT) 18, 20–21, 33, 71, 151, 180, 185
DNA wallet 119
documentary collection (DC) 243
documentation processes 227, 250
drugs
  availability of 232
  spurious 227, 231
  supply chain 231
  traceability 210
  trafficking in 233

Ebola epidemic 121
economic
  instability 67–68
  transaction 179
Ecosystem of Trust 252–253
effective resource management 169
efficiency 206, 210, 247
electronic health record (EHR) 120, 226, 228, 233; *see also* data analysis, healthcare
electronic payment system 151
electronic traceability system 142
emission trading system (ETS) 200
e-money 70
employee education 80
encryption 4–9, 25, 154
energy
  consumption 155, 161, 201–202
  efficiency 208
  grid 206–207
  markets 201
  trading agreements 206
  waste 38
enhanced efficiency 71, 247, 250–251
Enterprise Resource Planning (ERP) software 71, 185
environment 35, 41, 43–44
  blockchain 118
  for goods and products 36
  intelligent 117, 120–121
  multi-stakeholder 120
  public health surveillance and outbreak detection 121–122
  requirements 38
  social 60
  supply chain 91
  supportive 66
  sustainability 38, 40, 43
  tracking and traceability 121
errors 73, 98–99, 184, 240, 247–248, 250–251
Ethereum 155, 161–162
  blockchain architecture 90–91
  technology 200
Ethereum Address (EA) 135–136
ethical behaviour 133
ethnic disparities 89
eZTracker 231

FarmaTrust 231
fashion industry 200, 209–210
  ecosystem 207
  practical impact in 207
fault tolerance 158, 163, **163**
finality 161, 169

Index

final reachability matrix 56, 57
finance/financial 153–154, 240
    planning 251
    of trade 184–185
    transactions 185
fingerprints 115
FOMO 69
food
    industry 182
    products 178–179
    safety 181, 183
    security 186
    supply chain 43, 184, 248
    tracking 36
    waste 183
foodborne diseases 191
Food Standards Agency (FSA) 183–184
frauds 69, 152
    detection of 227
    insurance related 227
    management system 192
    probability of 246
    risk and protect against 243
"fully traceable" systems 129

Ganache framework 135
General Data Protection Regulation (GDPR) 120, 123
generic blockchain solution 188
genomics
    advancements in 120
    sequencing 229
global
    trade 185, 246–247, 251
    warming 217
gold
    illegal activities 133
    information 132
    market 130, 133
    mining industry 131
    provenance and legitimacy of 142
    shipment 139–140, *140*
    substandard 132
    supply chain 130–132
    tax clearance process 138, *139*
    trading 129, 136, 142
    transactions 133
GoldNet 133, 136–137
    backend development 135
    business processes 136, *137*
    frontend development 134–135
    high-level abstraction 134, *134*
    Prototype 134
    simulation environment 135, *135*
    smart contracts for 135
    supply chain 131–132, *131*
    system framework 134
    traceability 137
    workflow design 135–136
GoldNet smart contract design 136
    extraction and tokenisation 136–138, *138*
    sales 140–141, *140*
    shipment 139–140, *140*
    taxation clearance 138–139
    third-party tracking 141, *141*
gold supply chain 129–138, 142, 143
gold traceability
    description of 129–130
    future directions 143
    GoldNet *see* GoldNet
    literature review 130–133
    methodology 133
    policy implications 142–143
    practical implications 142
    research process 133, *134*
Good Manufacturing Practice (GMP) chain 94
Google's DeepMind healthcare AI 117
Graphene processors 77
green technology 203, 208–211, *208*; *see also* sustainability of green technology

Hadoop MapReduce 114
hash functions 5–9
HBase 114
healthcare sector, artificial intelligence and big data application in 116–117
    advancements 120
    clinical research and trial management 119–120
    consent and data ownership 117
    data monetization 117–118
    data sharing 120
    integration and collaboration 119
    intelligent environment 120–121
    interoperability 120
    for omnichannel healthcare 122–123
    patient centricity 117
    personalized 118

public health surveillance and
   outbreak detection 121–122
research and development 118–119
smart contracts 120
tracking and traceability 121
healthcare sector, blockchain
   applications in 240
   challenges in 225–227, 233–234
   consumers of 218
   corruption 226–227, 230–231
   data analysis *see* data analysis,
      healthcare
   demand of 219
   description of 217–219
   early diagnosis and treatment of
      life-threatening diseases 229
   efficiency in medical insurance
      segment 232–233
   electronic healthcare record 228
   improved efficiency in medical
      supply chain 231–232
   ineffective diagnosis of diseases 226
   interest and importance of 221–222
   medical service offerings and
      customer empowerment
      229–230
   methodology 219–220
   opportunities 227
   patient historical medical records
      225–226
healthcare services 153–154
   big data analytics in 105
   consumers of 223
   demand for 217
   inequity in 218
   organizations 106, 226
   providers 118
   supply chain 220
   system 218
Health Wizz 117–118
home healthcare service 217
Hong Kong Monetary Authority
   (HKMA) 247–248
human errors 247–248, 250–251
humanitarian supply chain process 210
Hyperledger Fabric blockchain 94,
   245, 247

Ideation Sprint workshop 166
identity management 209
immunization goals 89
immutability 5–9, 90, 119, 132, 179,
   218, 228, 246, 250–251
   of blockchain 183, 207
   functional attributes of 153
improvement planning 169, 171
increased efficiency 247
Industry 4.0 34–35, 49, 218
influenza vaccination 89
information and communication
   technology (ICT) 242–243
initial reachability matrix 56, **56**
inoculation chain 94
insurance claims 114, 120, 232–233
integrated medical records 226
integration, dimensions for 116–117
integrity 3, 5–6, 73, 95, 130, 132, 151,
   200, 208
intelligent healthcare environment
   120–121
intelligent vaccine management
   and supervision system
   (IVMSS) 91
intelligent vaccine supply management
   system (IVSCMS) 90,
   93–94, 100
intermediaries 5, 77, 117, 154–155,
   178–179, 184, 187, 208–209,
   232, 234
International Organization for
   Standardization (ISO) 181
international trade 243
   applications of 245
   Azarc's blockchain solution and
      ecosystem of trust 252–253
   and blockchain 242, 251–252
   challenges in 243
   complexity 250
   cost reduction 246–247
   customs clearance 242
   description of 240–242
   explosion of 242
   high cost of investment 248
   immutability of 246
   increased efficiency 247
   interoperability of 249
   lack of regulation 248
   management and theory 250–252
   practices 242
   scalability of 249
   security and privacy 249–250
   sustainability of 249
   traceability and transparency 246
   trade finance 243
Internet and Mobile Association of
   India 50

Internet of Things (IoT) 90, 105, 119, 142, 153–154, 186, 200–201
   Application Programming Interface 192
   incorporation of data 116
   integration 208
   monitoring 101
   for real-time monitoring 89, 91
   sensors 90–91, 93, 100
Internet of Transactions 5
interoperability 120, 156, 209, 249, 251
interpretive structural modelling (ISM) 52, 57–58, 58, 61–62
interrelationship 51–53, 60–61
intrinsic tamper-proof quality 180
investment, cost of 248
ISM-MICMAC approach 51

Keras Sequential API 97
key performance indicators (KPIs) 168, 170–171
knowledge
   management 202
   sharing, element of 207–208
Knowledge to Action (KTA) framework 168
know-your-customer (KYC) 70, 245
Kyoto Protocol 200

latency 160–161
   high 161
   lower 161
Layer 2 (L2) solutions 168–169
leadership 42
   motivation of 167
   support and regulatory 44–45, 80
learning techniques 202
letter of credit (LOC) 243, 245
Lifecode blockchain 228
life expectancy 217
literature 4–5
   academics/academy 234
   big data and artificial intelligence 106–107
   blockchain technology 52–53, 153
   data analysis, healthcare 220, 221
   gold traceability 130–133
   sustainability of green technology 199
   synergy 106–114

vaccine supply chain management 90–93
livestock sector, blockchain technology in 180–181, 184–191
   administration and enforcement systems 192
   certification and regulation 183–184
   customer feedback 186
   finance of trade 184–185
   quality and safety assurance 182–183
   smart contracts 184
   supply chain traceability 181–182
   unified connectivity 185–186
long short-term memory (LSTMs) 90, 91

machine learning (ML) 95, 118, 223
   Adaptive Boosting (AdaBoost) 96
   algorithms 90, 122
   analytics 93
   approaches 99
   deep convolutional neural network with pool layer 97
   deep neural network 97
   implementation of 89–90
   models 93
   multiple regression 96
   techniques 91
   temporal fusion transformer architecture 98, 98
   XGBoost framework 96–97
markets in crypto assets (MiCA) 70
mass balance systems 129
MCDM method 53, 62
mean absolute error (MAE) 98–99
mean absolute percentage error (MAPE) 94, 98–100
meat
   blockchain-enabled supply chain of 188, 189
   handling 182
   sector 180–181
meat value chain 180
MedChain 117
medical
   data storage 226
   devices 230
   fraternity 218, 230
   healthcare system 230
   history 226, 228–229, 232
   insurance 120, 232–233

records 117, 217–218, 222, 224–226
services 222, 226, 229–230
staff identification 121
supply chain 230–231, 233
treatment 229–230
MedRec 117
Metaverse 68
MICMAC 52–53, 57, 59–62, *59*
Micro, Small, and Medium Enterprises (MSMEs) 185
microgrids 201, 207
miners/mining 5, 7–8
    efficiency 77
    process 133
    types of 7
Monetary Authority of Singapore (MAS) 247–248
money laundering 69

Nanda Feeds Ltd 186
Nebula Genomics 119
net zero 199, 203, 206, 210
neural science 117, 119
Neurogress 119
non-repudiation 6
non-volatile data 58–59, 61
novel optimization strategies 169

omnichannel healthcare 122–123
on-chain coins 70
open account (OA) 243
organization 42–43, 79–80, 130–131, 151, 164–167, 170
    adoption, management in 66–67
    enablers of blockchain adoption in 75, *75*
    framework for 166–167
    illegal 129
    infrastructure 169
    innovation management 169
    preparedness 79
    readiness for blockchain 79
    self-assessment 168
organisational capabilities 66–81
organs
    trafficking 218, 227, 233
    transplant 226–227
outbreak detection 121–122

paper-based processes 241
paper-heavy tasks 246–247
paper-intensive procedures 245

"paper shuffling" processes 242
partitioning 57, 58, 60, 168–169
partnerships 71, 80
patients
    behavior 114
    centricity 117
    data management system 228
    diagnosis and treatment of 217–218
    history 232–233
    records 210
    tracking 118
payments
    methods 156
    systems 208
peer scrutiny 220
peer-to-peer (P2P)
    energy trading 206
    global transactions 37
    network 35, 159, 201, 209
    systems 199
    topology 33
    transactions 50
penetration
    academic 221–222
    degree of 223
    digital 234
Performance Expectancy of Blockchain 77
performance metrics 169
permission management systems 227
personalized healthcare 118
pharmaceutical industry 37
pharma supply chain 231–232
point of sale (PoS) 188
policy theory 251–252
Ponzi schemes 68, 70
population 89, 114, 178, 200, 217–218
Practical Byzantine Fault Tolerance (PBFT) 163
price volatility 69
PRISMA flow chart 202–203, *204*
privacy 9, 18, 37, 115, 119, 154, 156, 217–218, 232, 249–250
private blockchain systems 73, 159–161, **159**
private key management 250
Process, Institutional, Market, Technology (PIMT) framework 168, 170
process efficiency 247
proof-of-concept systems 245

proof-of-stake (PoS) 161, 163, 199, 244, 249
proof-of-work (PoW) 154–156, 161, 163, 199, 201–202, 244, 249
Prospect Theory 74
publications 20
    characterization of 3–4
    patterns 4
public blockchain 159, **159**, 227, 246, 249
    applications 162
    systems 161
public health 191
    needs 89
    surveillance 121–122

QR codes 189–191
quality 182–183
    accreditations 183
    patient care 114
quantum computing 77
quarantine restrictions 122
quick response (QR) 186, 206

racial disparities 89
React application 134
ready-to-eat (RTE) meat products 182
real-time
    data contribution 252–253
    monitoring 209
records
    integrity and safety of 228
    keeping systems 180
recurrent neural networks 90
regulation/regulatory
    compliance 80
    lack of 248
    reporting 136
    uncertainty 248
ReLU activation function 97
renewable energy
    generation 206–207
    production 207
Renewable Energy Certificates (RECs) 207
resource consumption 161
RFID tags 188, 190, 192
Ripple (XRP) 67
root mean square error (RMSE) 94, 98–100

SADC industrial strategy 142
Sankey diagram 17

scalability 155–156, 162, 194, 249
    addressing 162
    of blockchain systems 169
    challenges 162, *162*
    lack of 249
    trilemma 162
scientometric analysis 3–4, 9–10, 12, *12*, 24
SCOPUS database 220, 244
security 77, 119, 152, 154, 162, 181, 217–218, 249–250
    cryptography for 66
    enhanced 209
    potential 156
    risks 250
"security by design" principle 115
self-determination theory 74
service delivery 105, 234
SHA-256 hashing algorithm 7
Shanghai Yangshan Port 245, 247
shared distributed ledgers 247
simulation 101, 142
slaughterhouses 182–184, 188, 191
small- or medium-sized enterprises 79–80
small-scale producers 179
smart contracts 18, 68, 118, 120, 184, 198, 232, 245, 247–248
    on blockchain network 135–136
    for electricity usage 199
    function 136
    for GoldNet 135
    implementation of 94, 207
    systems 156
    use of 115
    utilization of 100
    violations of 190
smart metering systems 200–201
smartphones 188, 190–191
social
    environment 60
    learning theory 74
soil deterioration 178
solo-authorship 220
stakeholders 130–132, 179, 185–186, 193, 209–210, 228, 230, 243, 250
    acceptance 164
    in collaborative efforts 130
    data transparency and accessibility for 241
    industry 207

information flows through 72
internal and external 71–72
involvement 164
network 73
perception 58
potential use of blockchain technology 243
value chain 119
strategic management 80
Stream Computing 114
streamline business processes 210
structural self-interaction matrix (SSIM) 53, 56, **56**
supply chain 66–67, 70–73, 153–154, 178, 184, 240, 242, 247, 250
   aspects of 240
   collaboration and partnership 71
   complexities and susceptibilities inherent in 91
   data infrastructure and process efficiency 71
   environments 91
   friction within 187–188
   implementation 245
   integrity of 95
   of meat products *188*
   networks, transparency among 38
   regulatory compliance in 246
   strategies 130
   technical expertise 71
   traceability 181–182, 246
   traditional 187
   transactions in 94
   transparency 89
   visibility and resilience 246
supply chain management (SCM) 34, 41–42, 73, 131, 198, 210; *see also* supply chain
   benefits of blockchain to 45
   blockchain implementation for 42–43
   blockchain readiness model 40–45, *44*
   blockchain technology 34–39
   deployment for 34
   description of 33–34
   technology deployment in 39
sustainability 43, 156, 198, 208, 252
   evidence of 204–205
   of food supply chains 91
   Green Technology for 202–203

sustainability of green technology blockchain implementation 203
   conceptual framework 207–211
   data analysis and discussion 202–204
   description of 198
   features of blockchain technology 203–204
   knowledge management and capabilities of blockchain technology 201–202
   literature review 199
   PRISMA flow chart 202–203
   transforming features of blockchain technology 199–201
sustainable pilot projects 80
SWOT analysis 170
synergy
   artificial intelligence and big data application in healthcare sector *see* healthcare sector, artificial intelligence and big data application in
   data accuracy and lack of standards 116
   data security and privacy 115
   description of 105–106
   managerial implications 123
   research implications 123–124
   social implications 124
   study implications 123
   systematic literature review 106–114
   transparency and trust 115–116
system modeling 166

taxation clearance 138–139
technical infrastructure 80
technology 41–42, 199, 202–203, 209, 218–219, 252
   adoption of 77, **78**, 155–156
   advancement in 226–227
   benefits of 206
   comparative analysis of using 206
   end user perception of 40
   features 218–219
   in healthcare sector 222–223
   innovative 40
   performance 170–171
   potential applications of 153–154
   selection 167
   speed and efficiency of 69

Technology, Organization, and
    Environment (TOE)
    framework 34, 39–41, 44, *44*,
    163, 167–168
Technology Acceptance Model (TAM)
    34, 39–40, 167–168
technology readiness (TR) 34
TE-FOOD initiative 191–194
    application models 192–193
    end-to-end system 192
    extensive range of offerings 193
    supply chain 192, *192*
telecommunication 25, 247
temporal fusion transformer
    architecture (TFT) 98, *98*
    architecture 99
    forecasting 100–101
Theory of Diffusion of Innovations
    (DIT) 39
Theory of Planned Behaviour 74
Theory of Reasonable Action
    (TRA) 39
T-Mining 245
Token Forwarding Device (TFD)
    token 191
tokenisation 136, 199, 209
traceability 72, 118, 121, 178,
    181, 191, 209–210,
    246, 250–251
    adequate 182
    definition of 181
    of food items 193
    integration of 181
    loops 132
    supply chains 181–182, 246
    systems 181–182
trade finance 184–185, 240, 243,
    245–247
    distributed ledger technology in 248
    realm of 245, 247
    transaction 243
    types of 243
TradeLens 245
trading contracts 199
traditional payment methods 154–155
traditional supply chain (TSC)
    49–51, 53
    adoption of 60
    BCT integration in 58
    information sharing in 58
    information transfer across 51
    performance and ease of 59

throughput and latency issues in 59
transaction/transactional 7–8, 72,
    151–152, 160, 210, 248
    in blocks 129
    completion of 155
    confirmation 158
    efficiency of 68
    forms of 68
    gold 133
    history of 77
    latency 162
    nature of 161
    processes 153, 156, 161,
        168–169
    recording of 3, 161, 198
    in supply chain 94
    tamperproof record of 208
    throughput 162, 168–169
    traceable 207
    transparency 151, 207
    verification 72, *72*
transparency 52–53, 77, 118, 152, 154,
    181, 191, 199, 206, 218, 246,
    250–251
    blockchain technology in livestock
        sector *see* livestock sector,
        blockchain technology in
    of data on blockchain 246
    degree of 180
    description of 178–181
    TE-FOOD initiative 191–193
transparent billing 208
transportation process 190
trialability 163–164
trust 132, 181, 208
    and authenticity 38
    in clinical trial data 118–119
    cryptographic evidence of 116
    degree of 180
    via auditability 121
"trustless" network 5
TypeScript programming language 134

uncertainty 62, 67, 187–188, 248
unified connectivity 185–186
Unified Theory of Acceptance and
    Use of Technology (UTAUT)
    39–41, 74–76, **74**, **76**, 168
United States Sentencing Commission
    (USSC) 240
University Health Network (UHN) of
    Canada 119

US Securities and Exchange Commission (SEC) 67
utility token 70

VaccCoin virtual currency 94–95
vaccine supply chain management 101
    AI technologies in 90
    blockchain technology for enhanced 92, 94–95
    deficiencies in 89
    demand data 95
    description of 89–90
    evaluation metrics 98–99
    implementation of 90
    literature review 90–93
    machine learning experiments *see* machine learning experiments
    methodology 93–94, *93*
    pricing and coordination 90
    quality assurance in 91
    results and discussion 99–101, **100**
    system architecture 95, *95*
    tamper-proof tracking of 100
vaccine/vaccination 89
    authenticity and safety 91
    demand forecasting 90
    production and expiry dates of 94
    traceability 94
    wastage 100
value-adding service 201
venture capital allocation 69
Verathread middleware 252
viral influenza pneumonia 122
virtual currency 66, 94
Vitalik Buterin 155, 170
VOSviewer 9–11
vulnerabilities 93

Walmart 183, 245
waste treatment management 36
Web3JS Software Development Kit (SDK) 135
Web of Science (WOS) database 3, 10–12, 40
Webpack 134
word cloud analysis 224–225, *224*
work culture 166, 169
World Health Organization (WHO) 116, 217, 227
world population 217
World Trade Organization (WTO) 181
World Wide Web 68

XGBoost framework 96–97, 99